Compilation of

Radiological Physics Examinations (RAPHEX)

Vol. 1 Therapy

SUSAN V. BROWNIE, *Editor-in-Chief*
SIAMAK SHAHABI, *Technical Editor*

ADVANCED MEDICAL PUBLISHING
Madison, Wisconsin
1994

Published by
Advanced Medical Publishing
5331 Brody Drive, Ste. 201
Madison, WI 53705

ISBN: 1-883526-00-0

Preface

Since 1970 Raphex exams have been prepared by members of RAMPS, the New York chapter of the American Association of Physicists in Medicine. The purpose of the exam is to help residents in diagnostic and therapeutic radiology review their knowledge of radiological physics.

This volume is a compilation of questions from 1984 to 1992. The questions have been divided into topics to make the review process easier.

The answers are accompanied by brief explanations, along with any appropriate calculations; however, residents are urged to discuss these problems further with their physics faculty.

All questions have been reviewed and many revised and updated. We appreciate any comments or corrections you may have. These can be sent to:

Susan V. Brownie
c/o Advanced Medical Publishing
5331 Brody Drive Ste. 201
Madison, WI 53705

Editors

Susan Brownie, M.S.   *New York University Medical Center, New York*
Doracy Fontenla, Ph.D.   *Montefiore Medical Center, New York*
Richard Riley, Ph.D.   *Northshore University Hospital, New York*
Howard Amols, Ph.D.   *Columbia University, New York*
Lawrence Reinstein, Ph.D.   *S.U.N.Y. at Stony Brook, New York*
James Galvin, Ph.D.   *New York University Medical Center, New York*
Edward Nickoloff, Ph.D.   *Columbia University, New York*
Maynard High, Ph.D.   *New York Medical College, New York*
Lawrence Rothenberg, Ph.D.   *Memorial Sloan-Kettering Cancer Center, New York*
Alan Schoenfeld, M.S.   *Montefiore Medical Center, New York*
James Summers, M.S.   *Winthrop University Hospital, New York*

GENERAL

Questions
&
Answers

GENERAL Contents

1. Radiological Units

G1.   Which of the following is **not** equal to a centigray?

A. 100 erg/gram

B. 1 rad

C. 1/0.873 roentgen absorbed in <u>air</u>

D. 1/0.873 roentgen absorbed in tissue

G2.   The quality factor (Q) in radiation protection is most closely related to:

A. roentgen to cGy conversion factor

B. half-value layer

C. electron equilibrium

D. mass attenuation coefficient

E. relative biological effectiveness

G3.   The difference between exposure and dose is the difference between:

A. the rad and the gray

B. absorption of ionizing radiation and biological effect

C. photons and charged particles

D. ionization in air and absorption in a medium

E. ionizing and non-ionizing radiation

G4.   The energy absorbed by a mass of air from x- or γ-rays per roentgen is:

A. greatest in the photoelectric region

B. lowest in the Compton region

C. greatest in the pair production region

D. dependent on the temperature and pressure

E. the same, regardless of the x- or γ-ray energy

G5.   The f-factor:

A. is the roentgen to cGy conversion factor

B. is greatest for high Z materials and low energy photons

C. is 0.873 in air

D. all of the above

E. none of the above

G6-12.   Match the unit with the quantity it measures (answers may be used more than once or not at all).
A. dose equivalent
B. exposure
C. absorbed dose
D. activity
E. energy

G6.   Gray   *absorbed dose*   C
G7.   Becquerel   *activity*   D
G8.   Rem   *dose equivalent*   A
G9.   Rad   *absorbed dose*   C
G10.   Sievert   *dose equivalent*   A
G11.   Curie   *activity*   D
G12.   Roentgen   *exposure*   B

G13-16.   Match the unit with the quantity it measures (answers may be used more than once or not at all).
A. frequency
B. wavelength
C. power
D. absorbed dose
E. energy

G13.   Electron volt   *energy*   E
G14.   Hertz   *frequency*   A
G15.   Joule   *energy*   E
G16.   Watt   *power*   C

G17.   Exposure is:
A. the amount of energy in joules/kg transferred from a photon beam to a material
B. only defined for photons and charged particles below 3 MeV
C. the charge liberated by photons in a given mass of air
D. the absorbed dose multiplied by the quality factor
E. none of the above

G18-21.   Match the quality factor (Q) used in radiation protection with the type of radiation:
A. 10
B. 2
C. 1
D. 0.693
E. 20

G18.   1.25 MeV gammas   C
G19.   100 keV x-rays   C
G20.   Fast neutrons   A E
G21. ^{99}Mo betas   C

G22.   1 mSv is equivalent to:
         A. 1 mrem
         B. 10 mrad $Sv = 100 rem$
         C. 100 mroentgen
         D. 10 mCi
         E. 100 mrem

G23.   Dose equivalent is greater than absorbed dose for:
         A. x-rays above 10 MeV
         B. superficial x-rays
         C. electrons
         D. neutrons $QF = 20$
         E. all charged particles

G24.   Which of the following is *not* true?
         A. 100 MHz = 10^8 cycles/sec $Hz = 1 cycle/sec$ $f \times 10^{-15}$
         B. 1 curie = 3.7×10^{10} Bq $p \times 10^{-12}$
         C. The speed of light is 3×10^8 m/sec $\checkmark$ $n \times 10^{-9}$
         D. c = wavelength/frequency $c = \lambda f$ $\mu \times 10^{-6}$
                  $c = \lambda \cdot freq$ $m \times 10^{-3}$
                                                                  $M \times 10^{3}$

G25.   A given exposure:
         A. always results in the same absorbed dose to muscle or bone   *only same in air*
         B. is a measure of the ability of a photon beam to ionize air $\checkmark$
         C. is a measure of the ability of a particle beam to ionize air
         D. can be measured in roentgen in the SI system
         E. all of the above

G26.   Which of the following is true? For x- or gamma radiation, at standard temperature and
         pressure, a roentgen is equal to:
         A. one electrostatic unit (esu) of charge per cc of air $cc = cm^3$
         B. 2.58×10^{-4} coulombs/kg in air
         C. an absorbed dose of 0.873 cGy in air $\checkmark$
         D. A, B, C
         E. B, C only

G27.   One patient received an exposure of 4 roentgens to an area 10 × 10 cm, while a second   *total mg*
         received 1 roentgen to an area 20 × 20 cm. The absorbed dose to the second patient would *imparted to*
         be: *tissue irrad'd*
         A. less   *dose = energy/kg*   *but integral dose*
         B. the same   *is ↑ for ↑ F.S.*
         C. more
                  exposure is less - irrelevant of F.S.
G28.   The kerma is the: $D = fx$
         A. energy per unit mass absorbed or retained along the path of a charged particle
         B. energy per unit mass transferred from charged particles
         C. energy per unit mass transferred from photons or uncharged particles to charged particles
         D. charge released by photons as they pass through a specified amount of air

G29-32.   Match the following:
A. 100 gray
B. 0.01 gray
C. 100 rad
D. rad × Q
E. gray × Q

1 gy = 100 rad

G29.   Rad   *.01 gray (B)*
G30.   Rem   *Rad × Q (D)*
G31.   Gray   *100 rad (C)*
G32.   Sievert   *gray × Q (E)*

G33.   The unit of exposure was *originally* defined as:
A. 1 R = 1 esu of charge per 0.001293 cm^3 of air at STP
B. 1 R = 1 esu of charge per m^3 of air at STP
C. 1 R = 1 C of charge per cm^3 of air at STP
D. 1 R = 1 C of charge per 1.293 kg of air
E. 1 R = 2.58 × 10^{-4} C of charge per kg of air   *← current definition*

esu = measured Ch
1 coulomb = charge

G34.   If 5 coulombs flow through a wire in 2 seconds, the current is:
A. 2 amps
B. 2.5 amps
C. 5 amps
D. 10 amps

current = coulombs/sec = charge/time
5/2 = 2.5

G35.   Which of the following is *not* an SI unit?
A. meter
B. kilogram
C. second
D. rad   *→ should use gray*
E. becquerel

2. Atomic Structure

2.1

G36-37.   Match the following characteristics of massive particles and/or electromagnetic radiations:
(answers may be used more than once).

	Charge	Rest Mass
A.	+1	0.511 MeV *positron*
B.	+1	about 930 MeV *proton*
C.	0	0 *photon neutrino antineutrino*
D.	0	about 930 MeV *neutron*
E.	-1	0.511 MeV *negatron*

G36.   Loses the most energy per unit path length. *B*
G37.   Results from pair production and eventually undergoes annihilation. *A*

G38-42. An electron is electrostatically _____ by a/an:
         A. attracted
         B. not affected
         C. repelled

G38.   Alpha particle   ~~B~~ A
G39.   Electron   C
G40.   Positron   A
G41.   Proton   A
G42.   Neutron   B

G43-46. Match the property to the particle:
         A. no charge; negligible mass
         B. no charge; mass of 1.6×10^{-27} kg
         C. charge of +1; mass of 9.1×10^{-31} kg
         D. no charge; mass of 9.1×10^{-31} kg
         E. charge of 2; mass of 4 amu

G43.   Neutrino   A
G44.   Alpha particle   E
G45.   Positron   C
G46.   Neutron   B

G47.   The rest mass of an electron is:
         A. 981 MeV
         B. 1.02 MeV
         C. 0.51 MeV
         D. 1 amu
         E. 0.51 keV

G48-51. Match the particle with the description.
         A. neutron
         B. proton
         C. antineutrino
         D. positron
         E. alpha $^4_2$He

G48.   Nucleus of a hydrogen atom B ^{1}H
G49.   Is emitted during pure beta minus decay C
G50.   Is created during pair production D
G51.   Initiates fission in ^{235}U A

G52.   Directly ionizing radiations do *not* include: *must be charged*
         A. electrons
         B. positrons
         C. neutrons → H+, in fat
         D. alpha particles
         E. beta-rays β particle

G53-57.   Give the charge carried by each of the following:
           A. +4
           B. +2
           C. +1
           D. 0
           E. -1

G53.   Alpha particle   *B*
G54.   Neutron   *D*
G55.   Electron   *E*
G56.   Positron   *C*
G57.   Photon   *D*

G58-61.   Match the following: (answers may be used more than once)
           A. electrons
           B. protons $p \rightarrow \beta^+ + \bar{n}$
           C. neutrons
           D. neutrinos $n \rightarrow \beta^- + p$
           E. gamma-rays

G58.   *E*   Most responsible for nuclear medicine imaging
G59.   *B*   Most responsible for MR imaging
G60.   *D*   Most difficult to detect
G61.   *A*   Emitted in beta minus decay

G62.   A neutron is heavier than an electron. The ratio of their masses is approximately:
           A. 10:1
           B. 500:1
           C. 1000:1
           D. 1400:1
           E. 1800:1

G63.   An electron, proton, and photon each have 1000 MeV total energy (kinetic energy + rest
           mass energy). Which of the following statements is true?
→   A. the electron travels at almost the speed of light
           B. the proton travels at almost the speed of light
           C. the photon travels at almost the speed of light
           D. the proton has the most kinetic energy
           E. the electron and the proton have the same rest mass

$E = mc^2$

$\dfrac{1000}{m} = c$ → photon does travel @ the speed of light

$t \xi \approx$ rest $Nrg + KA$

                .511   931

G64.   Which of the following is never emitted during radioactive decay?
           A. alpha particle
           B. proton
           C. positron
           D. gamma-ray
           E. neutrino

2.2.

G65.   Regarding isotopes:
A. elements can only have one stable isotope
B. elements can only have one radioactive isotope
C. isotopes of an element have the same number of protons but different numbers of neutrons
D. isotopes of an element appear in adjacent columns of the periodic table
E. all of the above are true

G66. ^{131}Iodine and ^{125}Iodine have:
A. different chemical properties
B. different Z values
C. occupy different columns on the periodic table
D. the same number of neutrons
E. none of the above

G67.   Elements which have the same Z but different A are called:
A. isobars
B. isomers
C. isotones
D. isotopes

G68.   Which of the following have different mass numbers?
A. isomers
B. isobars
C. isotones
D. all of the above

G69.   A certain radionuclide can decay by either beta minus emission or positron emission. The two daughter nuclei are:
A. isomers
B. isobars
C. isotones
D. isotopes

2.3.

G70.   The mass number (A) of an atom is equal to the sum of the:
A. neutrons
B. protons
C. nucleons
D. atomic masses minus the total binding energy
E. atomic masses plus the total binding energy

G71.   The number of neutrons in a cobalt-60 atom (Z = 27) is:
        A. 27
        B. 60   *60 − 27 = 33*
        C. 33
        D. 7
        E. cannot tell from information given

G72.   Tritium is an isotope of hydrogen with the symbol ^{3}H.
        A. its atomic number is 3
        B. its mass number is 3
        C. its atomic number is 2
        D. its mass number is 1
        E. it has 3 neutrons

G73.   The number of electrons in a neutral atom equals the:
        A. mass number
        B. atomic weight
        C. atomic number
        D. nucleon number
        E. valence number

G74.   The energy equivalent of one atomic mass unit is approximately:
        A. 10^0 eV
        B. 10^3 eV
        C. 10^6 eV
        D. 10^9 eV
        E. 10^{12} eV

        $1 \, AMU \cong 931 \, MeV = .931 \times 10^3 \times 10^6 = .9 \times 10^9$

G75.   The number of electrons per gram of material is approximately equal to (N_A = Avagadro's number):
        A. $N_A Z$
        B. $N_A A$
        C. $N_A A/Z$ *$\dfrac{Z N_A}{A}$*
        D. $N_A Z/A$
        E. N_A

2.4.

G76.   Electron binding energy:
        A. is greater in the K-shell than the L-shell
        B. is greater for the K-shell of barium than the K-shell of hydrogen
        C. increases with increasing Z
        D. all of the above
        E. none of the above

G77.   The binding energy of an electron in the K-shell is:
        A. the energy the electron needs to stay in the K-shell
        B. the energy needed to make a transition to the L-shell from the K-shell
        C. the energy needed for an electron to jump from the L- to the K-shell
        D. the energy needed to remove an electron from the K-shell
        E. none of the above

G78.   The K-shell binding energy of tungsten (Z = 74) is approximately 69.5 keV. Therefore, one
        would expect the K-shell binding energy of oxygen (Z = 8) to be about:
        A. 92 keV
        B. 9.2 keV
        C. 0.81 keV
        D. 3 eV

        Binding energies $\propto Z^2$

        $74^2 / 8^2 = \dfrac{69.5}{E}$

G79.   Which of the following is **not** true? The electron binding energy:
        A. decreases with increasing distance from the nucleus
        B. decreases with increasing nuclear charge
        C. is a few electron volts for the outer electrons of an atom
        D. must be overcome if ionization is to take place

        [B.E $\uparrow$ as $Z^2 \uparrow$]

G80.   When an electron is removed from an atom, the atom is said to be:
        A. radioactive
        B. ionized
        C. inert
        D. excited
        E. metastable

G81.   From the information below calculate the minimum energy required to separate a deuteron into
        its component particles: [1 amu = 931.2 MeV]

Particle:	Amu:
proton	1.00727
neutron	1.00866
deuteron	2.01355

        A. 2.02 MeV
        B. 1.875 MeV
        C. 2.22 MeV
        D. 2.38 MeV
        E. 4.03 MeV

        Deuteron ^2H

        .00238
        931.2

        1.00727
        1.00866
        2.01593
        − 2.01355
        .00238

        476
        2380
        71400
        142000
        2216256

G82.   An atom with an ionization potential of 13.4 eV is bombarded with 7 eV photons. The
        minimum number of photons needed to ionize the atom is:
        A. 1
        B. 1.91
        C. 2
        D. 3
        E. none of the above

        need to have $\uparrow E$ photons
        of @ least 13.4 eV

G83.   The binding energy per nucleon usually: *BE/A* *increases*
        A. does not change in beta decay
        (B) increases after radioactive decay to the ground state
        C. is independent of Z
        D. is minimum for intermediate values of Z

G84.   Heavy nuclei contain more neutrons than protons because:
        (A) they are needed to counteract the repulsion between protons
        B. the universe contains more neutrons than protons
        C. protons become radioactive
        D. none of the above

G85.   The binding energy per nucleon is:
        A. equal to the atomic mass
        (B) larger for stable nuclei than for radioactive nuclei
        C. equal for different isotopes of the same element
        D. decreased when an atom is ionized *not*

2.5.

G86.   The maximum number of electrons in a shell:
        A. is always 8
        B. is always 2
        (C) is $2n^2$, where n is the principle quantum number
        D. none of the above

G87.   The maximum number of electrons in the outer shell of an atom is:
        A. $2n^2$
        (B) 8 ——————→ *b/c of eight groups in periodic table.*
        C. 16
        D. 32
        E. 2

        does not count for inner shells which follow $2n^2$ rule

G88.   For the elements Z = 16, 17, 18, and 19 (sulfur, chlorine, argon, and potassium), how many
        have 8 electrons in the M-shell?
        A. 0
        B. 1
        (C) 2
        D. 3
        E. 4

K 2
L 8
M 18

Z	K:L	leftover
16	10	6
17	10	7
(18)	10	8
(19)	10	9

only possible b/c have 8 or more for M shell.

3. Nuclear Decay

3.1.

G89. Which graph below approximately represents radioactive decay?

A

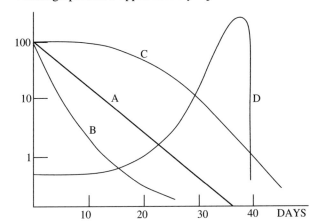

G90. Activity is defined as:

A. dN/dt
B. $0.693/T_{1/2}$
C. 0.693/HVL
D. $0.693/\lambda$
E. $1.44 \times T_{1/2}$

N = number of atoms
t = time
T = half-life
HVL = half-value layer
λ = disintegration constant

$A = \lambda N$

G91. During nuclear decay, energetic particles are emitted. The maximum energy of these particles is related to the concept of:

A. annihilation radiation
B. neutron capture
C. the exclusion principle
D. mass defect
E. none of the above

$A = A_0 e^{-\lambda t} = A_0 e^{-\frac{.693 t}{T_{1/2}}}$

$e^{-\lambda t}$

G92. Which of the following is **not** true?

A. exponential attenuation accounts for the fact that x-ray beams do not have a range
B. when a source decays exponentially, after 3 half-lives 12.5% of the initial activity remains
C. a radioisotope with a 14 day half-life will have more than 75% of its initial activity left after 7 days
D. for any one nucleus the actual time of decay is not known, only the probability of decay
E. all of the above, since none of them are true

G93. The half-life of a radioisotope is:

A. influenced by temperature and pressure
B. directly proportional to the decay constant
C. usually shorter for beta minus than beta plus emitters
D. less than the average life $\approx 1.44\, T_{1/2}$
E. all of the above

$\lambda = \frac{.693}{T_{1/2}}$ $T_{eff} = 1.44 \cdot T_{1/2}$

$= \frac{1.44\,(.693)}{\lambda}$

$= \frac{1}{\lambda}$

G94. A source has a half-life of 12 hrs, and an initial activity of 10 mCi. After 3 days its activity will be _____ mCi.
A. 1.67
B. 3.33
C. 1.56
D. 0.31
E. 0.16

$$X = 10 e^{-\left(\frac{.693}{12}\right)72}$$

$$= 10\left(\frac{1}{2}\right)^6$$

G95. After 10 half-lives, the fraction of activity remaining in a source is:
A. $(1/10)^2$
B. 1/10
C. dependent on the initial activity
D. $(1/2)^{10}$
E. 9/10

$$\frac{A}{A_0} = e^{-\frac{.693}{10}}$$

G96. If a radionuclide decays at 1% per hour, about how long will it take to decay to 1/2 its original activity?
A. 10 hrs
B. 30 days
C. 50 hrs
D. 70 hrs
E. 90 days

$$\lambda = .01$$
$$\lambda = \frac{.693}{T_{1/2}}$$
$$\left(\frac{.01}{.693}\right)^{-1} = T_{1/2}$$

↓ Half-life

G97. The decay constant is:
A. inversely proportional to the half-life ✓
B. the fractional decay in a given time
C. equal to the reciprocal of the average life
D. all of the above
E. none of the above

$$\frac{1}{1.44} = .693$$

Average life = $1.44\, T_{1/2}$
$$= \frac{1}{\lambda}$$

G98. A ^{99m}Tc source (half-life = 6 hours) of 20 mCi will decay in 3 hours to about _____ mCi:
A. 17
B. 15
C. 14
D. 10
E. 5

$$X = 20 e^{-\frac{.693}{6}(3)}$$

G99. If an administered radionuclide undergoes biological elimination, the effective half-life is _____ the physical half-life.
A. longer than
B. equal to
C. shorter than

$$\lambda_{eff} = \lambda_{BIOL} + \lambda_{phy}$$

$$\frac{1}{T_{eff}} = \frac{1}{T_{BIO}} + \frac{1}{T_{phy}}$$ ✓

$$\frac{1}{T_{eff}} = \frac{T_{phy}}{T_{BIO} * T_{phy}} + \frac{T_{BIO}}{T_{BIO} * T_{phy}}$$

$$\frac{1}{T_{eff}} = \left(\frac{T_{phy} + T_{BIO}}{T_{BIO} * T_{phy}}\right)^{-1}$$

G100. Comparing two gamma-emitting radionuclides administered by the same route, for the same number of millicuries and the same biological half-life, the one with the shorter half-life will generally:
A. emit more gamma-rays per second
B. give a higher patient dose
C. give a lower patient dose
D. give the same patient dose

Total dose = Dose rate × average life
= A_0 (1.44) $T_{1/2}$

G101. The decay constant for ^{182}Ta is 0.006 day^{-1}. What is its half-life?
A. 167 days
B. 115 days
C. 0.0087 days
D. 0.006 days
E. 83 days

$T_{1/2} = \frac{.693}{.006 \, days} = \underline{\quad} days$

G102. If the biological and physical half-lives of a radioisotope are both 2 hours, the effective half-life is:
A. 0.25 hours
B. 0.5 hours
C. 1 hour
D. 2 hours
E. 4 hours

$\frac{2 \times 2}{2 + 2} = \frac{4}{4} = 1 \, hrs$

3.2.

G103. Which of the following is true of alpha decay?
A. A changes by 2
B. Z changes by 4
C. charge is not conserved
D. this is most likely in atoms with A < 82
E. this is most likely in atoms with Z > 82

[$^{4}_{2}$He is α w/o e^-]

G104. ^{226}Ra decays to ^{222}Rn by _____ decay.
A. beta minus
B. alpha
C. gamma
D. beta plus
E. B and C

γ decays seen later in the chain

G105. ^{222}Radon decays to ^{118}Po ("radium A") by:
A. alpha emission only
B. beta followed by alpha
C. alpha followed by beta
D. alpha plus electron capture

G106-109. Match the mode of decay and the change in atomic number (Z) (answers may be used more
than once).

A. $Z \rightarrow Z + 2$
B. $Z \rightarrow Z + 1$
C. $Z \rightarrow Z$
D. $Z \rightarrow Z - 1$
E. $Z \rightarrow Z - 2$

G106.   Alpha (α)   E
G107.   Beta minus (β^-)   B   } isobaric decay
G108.   Beta plus (β^+)   D   }
G109.   Isomeric   C

$p \rightarrow \beta^+ + neutrino$

$n \rightarrow \beta^- + p$

G110.   The mass number (A) changes only for _____ decay.
A. alpha (α)
B. beta minus (β^-)
C. beta plus (β^+)
D. electron capture
E. isomeric

G111-112. This diagram refers to the next two questions.

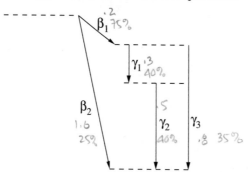

	Energy (MeV)	Abundance
β_1	0.2	75%
β_2	1.0	25%
γ_1	0.3	40%
γ_2	0.5	40%
γ_3	0.8	35%

G111.   The total energy emitted in a single disintegration in the decay scheme above is ____ MeV.
A. 0.2
B. 0.33
C. 0.5
D. 0.8
E. 1.0

G112.   In the decay scheme shown in the diagram, the total energy emitted in the pathway that
includes β_1 and γ_3 is divided among:
A. two gamma-rays
B. two gamma-rays and one beta-ray
C. one gamma-ray, one beta-ray and one antineutrino
D. one beta-ray and one antineutrino
E. two beta-rays and three gamma-rays

G113-118.   For the decay scheme below (answer A for true and B for false):

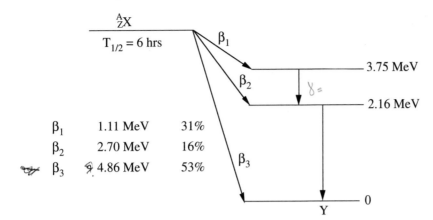

G113.   B   The mass number of Y is A + 1.
G114.   B   The atomic number of Y is Z – 1.
G115.   A   The decay of X is accompanied by the emission of antineutrinos.
A G116.   A   0.511 MeV beta-rays are emitted. b/c of continuous spectrum.
G117.   B   7.6 MeV beta-rays are emitted.
G118.   A   In any one disintegration, a total of 4.86 MeV will be emitted.

G119.   Two nuclides have the following properties:

	Nuclide I	Nuclide II
Atomic number	Z	Z – 1
Mass number	A	A
Atomic mass	M (MeV)	M – 2 (MeV)

Nuclide I may transform into Nuclide II by (answer A for true and B for false):
1. beta plus decay → needs @ least 1.022 MeV
2. isobaric transition
3. isomeric transition   no z △

A. 1 only
B. 1, 2
C. 2, 3
D. 3 only
E. 1, 2, 3

G120.   When $^{32}_{15}P$ decays to $^{32}_{16}S$:   β⁻
A. this is an example of positron emission
B. all the beta particles are emitted with the same energy
C. this is an isomeric transition
D. an antineutrino is emitted
E. none of the above

G121.   In the decay of ^{60}Co (Z = 27) to ^{60}Ni (Z = 28), _____ are emitted.
A. monoenergetic electrons only
B. monoenergetic positrons only
C. monoenergetic photons and electrons with a spectrum of energies
D. monoenergetic electrons and photons with a spectrum of energies
E. monoenergetic photons and monoenergetic electrons

G122.   A radionuclide with an excess of neutrons generally decays by:
A. alpha
B. beta minus
C. beta plus
D. electron capture
E. internal conversion

$n \rightarrow p + e^-$
$\beta-$

G123.   Which physical process emits a continuous spectrum of radiation?
A. alpha decay
B. isomeric transition
C. electron capture
D. beta decay
E. both A and D

G124. $$^{99}\text{Mo} \rightarrow {}^{99m}\text{Tc} \rightarrow {}^{99}\text{Tc}$$
$$\text{(I)} \qquad \text{(II)}$$

140 keV γ released and used for nuclear imaging

In the above decay of molybdenum, the modes of decay labeled (I) and (II) are, respectively:
A. beta minus, isomeric transition
B. isomeric transition, beta minus
C. beta plus, isomeric transition
D. beta minus, beta minus
E. electron capture, beta minus

G125-129. Answer the following questions based on the figure below.

A. always occurs
B. sometimes occurs
C. never occurs
D. cannot be determined from the diagram

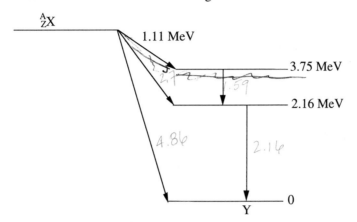

G125. B 1.59 MeV gamma-ray
G126. D 40 keV characteristic x-ray
G127. B 1.12 MeV beta minus   will get y up to 2.16 or 0 mev — the remainder of
G128. B 4.80 MeV antineutrino   E goes to antineutrino
G129. C 3.75 MeV gamma-ray

G130.   In the decay scheme shown in the diagram:

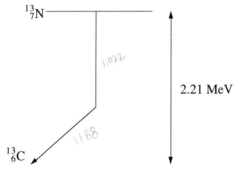

(A) the maximum energy of the positrons is 1.19 MeV
B. positrons are the only particles emitted ∅
C. the nuclear masses of ^{13}N and ^{13}C are equal ∅
D. the maximum energy of the positrons is 2.21 MeV ∅
E. the atomic number increases by 1 ∅

G131.   Positrons eventually combine with electrons in matter resulting in:
A. a particle with equivalent mass but no charge
B. equivalent energy in the form of a photon
C. two particles emitted in opposite directions
(D) two photons of equal energy emitted in opposite directions

2 γs of equal energy = .511

{ .511
{ .511

G132. A radioisotope decays to a nuclide with the same A, and with Z reduced by 1. The decay is:
A. alpha
B. beta plus
C. beta minus
D. electron capture
E. B or D

G133. Positron emission:
A. is sometimes accompanied by neutrino emission
B. cannot occur unless the energy levels of the parent and daughter nuclei differ by ≥ 511 keV
C. is followed by the creation of 511 keV photons
D. consists of monoenergetic positrons
E. all of the above

G134. Compared to ^{99}Tc, the total mass-energy equivalent of ^{99m}Tc is:
A. larger
B. smaller
C. identical

G135. A photon is detected outside a lead container with 5 mCi of ^{99m}Tc inside. The photon could:
1. have an energy of 130 keV
2. be an x-ray
3. be Cerenkov radiation
4. be an Auger x-ray
5. be annihilation radiation

A. 1 only
B. 1, 2
C. 3, 4
D. 1, 3, 5
E. 1, 2, 3, 4, 5

G136. Which massive particle and/or electromagnetic radiation results from a nuclear transition where the atomic number (Z) remains unchanged?

	Charge	Rest Mass
A.	+1	0.511 MeV
B.	+1	about 930 MeV
C.	0	0
D.	0	about 930 MeV
E.	-1	0.511 MeV

G137. A photon may be emitted following a/an _____ decay.
1. beta minus
2. beta plus
3. isomeric transition
4. electron capture

A. 1 only
B. 3 only
C. 2, 4
D. 1, 2
E. 1, 2, 3, 4

G138.   In decay by isomeric transition:
          A. the energy remains the same
          B. the atomic number Z decreases by 1
          C. the mass number A decreases by 1
          D. only gamma-rays are emitted → or as conversion electrons resulting
          E. A and Z remain the same   in char x-rays or Auger e⁻

G139.   Which massive particle and/or electromagnetic radiation carries off most of the energy
          when a nucleus decays by electron capture to the ground state?

	Charge	Rest Mass
A.	+1	0.511 MeV
B.	+1	about 930 MeV
C.	0	0 characteristic x-ray
D.	0	about 930 MeV
E.	-1	0.511 MeV

G140.   Select the change in atomic number (Z) which occurs during electron capture.
          A. Z → Z + 2
          B. Z → Z + 1
          C. Z → Z
          D. Z → Z – 1
          E. Z → Z – 2

G141.   Two nuclides have the following properties:

	Nuclide I	Nuclide II
Atomic number	Z	Z – 1
Mass number	A	A
Atomic mass	M (MeV)	M – 2 (MeV)

          Nuclide I may transform into nuclide II by:
          A. electron capture
          B. beta minus
          C. internal conversion
          D. alpha decay

G142.   The radioactive transformation below represents a/an _____ decay.

          $${}_{Z}^{A}X \rightarrow {}_{Z-1}^{A}Y + \gamma + \nu$$

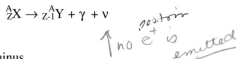

          positron
          ↑ no e⁺ is emitted

          A. alpha
          B. beta minus
          C. beta plus → need 1.02 or βr
          D. electron capture   ν = neutrino
          E. isomeric   γ = gamma-ray

G143-145. Match the process with the description:

A. a positron and an electron are created from a photon of energy greater than 1.022 MeV

B. electron + proton → neutron + neutrino   *e capture*

C. an electron and a characteristic x-ray are emitted

D. a positron and an electron combine to form two 0.511 MeV gammas

G143. B   K-capture   *electron capture ℮ + e → neutron neutrino*

G144. C   Internal conversion   *competes c γ*

G145. D   Annihilation radiation

G146.   Electron capture:

A. results in the emission of a neutrino

B. can compete with positron emission in isotopes with an excess of protons   *→ # of neutrons too low for stability*

C. results in characteristic x-ray emission   *x-ray product*

D. results in Auger electron emission   *competes c characteristic*

E. all of the above

in an EC electron (usually from K shell) combines w/ a proton to make a neutron then char x-ray and resulting Auger e⁻ emitted

G147.   In internal conversion: *→ an isometric transition (from the nucleus)*

A. Z and A remain the same

B. Z = +1, A remains the same

C. Z = -1, A remains the same

D. A = +1, Z remains the same

E. A = -1, Z remains the same

energy is directly transferred to an inner shell e⁻ (which is sometimes later emitted → also can emit γ ray)

G148.   Radionuclides which decay by internal conversion emit:

1. gamma-rays

2. characteristic x-rays

3. Auger electrons   *- b/c competes c characteristic x-rays*

4. beta minus

A. 1 only

B. 1, 3

C. 2, 4

D. 1, 2, 3, 4

E. 1, 2, 3

G149.   An alternative to the emission of characteristic radiation is:

A. internal conversion

B. K-capture

C. emission of an Auger electron

D. isomeric transition

EC, β⁺ & β⁻ no bone
IC - isometric

G150. Which of the following are *not* methods of radioisotope production used to prepare sources of medical radioisotopes?
A. separation from reactor fuel rods *Cs137 Sr90*
B. bombarding with protons in a cyclotron } *C11, N13, O14*
C. bombarding with deuterons in a cyclotron } *O15*
D. elution of a metastable daughter from a column containing the parent → *99mTc*
E. bombarding with neutrons in a cyclotron

cyclotrons can only accelerate charged particles!

charged particles

G151-155. Match the method of isotope production with the isotope.
A. sample bombarded with charged particles in a cyclotron *(usually short-lived β+ emitters)*
B. naturally occurring in uranium ore
C. sample placed in neutron flux of a reactor
D. separated from used reactor fuel rods
E. elution from the column of a generator

G151. A Oxygen-15
G152. B Radium-226
G153. E Technetium-99m
G154. D Strontium-90 *(usually used to monitor ion chamber constancy)*
G155. C Cobalt-60 *59Co bombarded w/ neutrons to make 60Co* *(occurs naturally)*

G156-159. Match the type of radioactive decay with the method of radionuclide production (answers may be used more than once):
A. alpha emission
B. beta minus emission
C. beta plus emission
D. isomeric transition

G156. B Bombarding a sample with neutrons in a reactor

G157. C Bombarding with protons in a cyclotron

G158. C Bombarding with deuterons in a cyclotron

G159. B A Separating fission products from reactor fuel rods

the ↑ the atomic # Z, the ↑ the n/p ratio, so when splitting elements like uranium, their are too many n, so β⁻ will likely occur

G160. Radionuclides used in positron emission tomography (PET) are usually made by:
A. bombarding nuclei with neutrons in a nuclear reactor
B. extraction from nuclear fuel rods
C. bombarding nuclei with protons from a cyclotron
D. extraction from uranium ore

on site cyclotron needed usually bc of short lived isotopes.

UNC does not have PET!

G161-165. Concerning radioactive equilibrium (answer A for true and B for false):

G161. B   Secular equilibrium occurs when the decay constant of the daughter is slightly greater than the decay constant of the parent.

G162. B   Transient equilibrium requires decay to a metastable state of the daughter.

G163. B   In transient equilibrium, the activity of the daughter is always less than that of the parent.

G164. A   Equilibrium may exist if the half-life of the daughter is shorter than that of the parent.

G165. A   Transient equilibrium exists if the half-life of the parent is somewhat greater than that of the daughter.

G166.   When cobalt-60 is created by placing cobalt-59 in the neutron flux of a reactor, the time taken to achieve more than 90% of the maximum possible activity is approximately:
A. 0.5 half-lives
B. 1 half-life
C. 4 half-lives
D. 10 half-lives
E. the value is unrelated to the half-life

$$A = (1 - .5^n) A_{max}$$

G167.   When equilibrium is established between a parent and daughter radionuclide:
A. the parent decays with the half-life of the daughter
B. the parent and daughter emit gammas of the same energy
C. the activity of the parent and daughter remain constant
D. the daughter decays with the half-life of the parent
E. the daughter always decays faster than the parent, as it has a shorter half-life

G168.   Which of the following occurs one month after a radium source (half-life 1600 years) is sealed in a tube with its daughter radon (half-life 3.8 days)?
A. transient equilibrium
B. secular equilibrium
C. equilibrium has not yet occurred

G169. ^{99}Mo (half-life 2.8 days) decays to ^{99m}Tc (half-life 6 hours). Which of the following has occurred 6 hours after the last milking?
A. transient equilibrium
B. secular equilibrium
C. equilibrium has not yet occurred

G170. Consider a parent/daughter radionuclide pair. Which of the following statements is true?
A. in transient equilibrium, the activities of parent and daughter are exactly equal
B. in secular equilibrium, the activities of parent and daughter are exactly equal
C. in secular equilibrium, the activity of the parent is greater than the activity of the daughter
D. in transient equilibrium the numbers of parent and daughter nuclei are equal
E. none of the above

G171. Radioactive equilibrium can occur only if:
A. the daughters' half-life is slightly less than the parents'
B. the daughters' half-life is much less than the parents'
C. the daughters' half-life is slightly greater than the parents'
D. the daughters' half-life is much greater than the parents'
E. A or B

G172. A certain naturally occurring radionuclide found in the soil has a half-life of 3 days. The total amount of this nuclide in the soil as compared to the amount which existed 6 days ago is:
A. 1/4
B. 1/2
C. very slightly less
D. very slightly greater
E. none of the above

3.5.

G173. 10 grams of a stable element are bombarded in a nuclear reactor, and the resultant activity is 40 Ci. The specific activity of this sample is:
A. 400 Bq/kg
B. 1.48×10^{14} Bq/kg
C. 0.25 g/Ci
D. 6.8×10^{-11} kg/Bq
E. 400 g/Ci

G174. A radioactive material has a disintegration rate of 7.4×10^6 becquerels (Bq). This is equal to _____ microcuries (μCi).
A. 5.0×10^{-1}
B. 2.0×10^2
C. 3.7×10^4
D. 7.4×10^6
E. 27.4×10^{10}

G175. 10 mCi = _____ MBq.
A. 3.7×10^{10}
B. 3.7×10^2
C. 2.7×10^{-11}
D. 2.7×10^5
E. 2.7×10^8

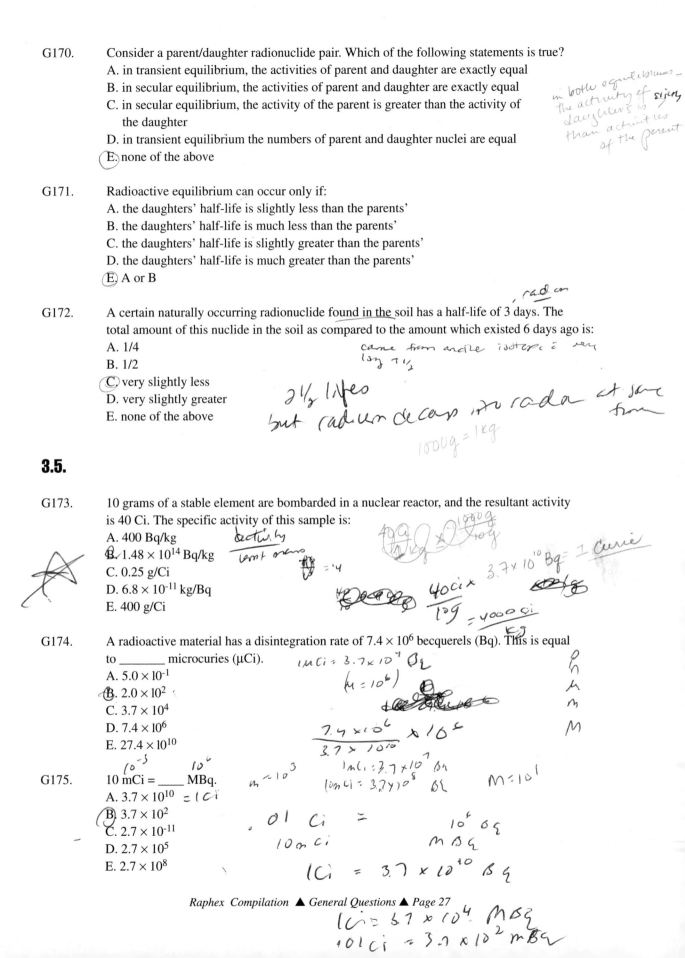

3.6.

G176.   A radiation worker standing 1 meter from a 5 mCi radioactive source with the following
         properties for 3 hours will be exposed to about ____ mR.
         [$\Gamma = 2$ (R-cm^2/mCi-hr) at 1 cm, $T_{1/2}= 60$ d, HVL= 0.03 mm Pb]
         A. 0.6
         B. 1
         C. 3
         D. 30
         E. 300

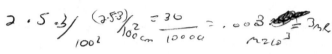

G177.   The exposure rate at 1 meter from a point source of 10 mCi of cesium-137 is:
         [Γ= exposure rate constant = 3.3 R-cm^2 /mCi-hr]
         A. 3 mR/hr
         B. 3.3 R/min
         C. 33 R/hr
         D. 3.3 R/hr
         E. 3.3 mR/hr

G178.   The exposure rate at 1 meter from 10 mg of radium is:
         A. 8.25 R/hr
         B. 82.5 R/min
         C. 8.25 mR/hr
         D. 0.825 R/hr
         E. 82.5 mR/hr

G179.   The reason that the gamma factor for cobalt-60 is greater than the gamma factor for cesium-
         137 is because:
         A. cobalt-60 has a shorter half-life
         B. cesium-137 has a higher energy gamma-ray
         C. cobalt-60 emits more than one gamma per disintegration
         D. A, B, and C all contribute
         E. none of the above because gamma for cesium-137 is higher than for cobalt-60

4. X-Ray Circuits

G180-183. This diagram of an x-ray circuit refers to the next three questions.

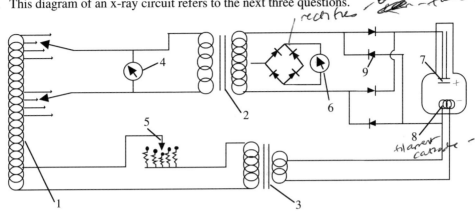

G180.   The mA meter is:
           A. 5
           B. 6
           C. 4
           D. 9

G181.   Which of the following is true?
           A. 7 is the anode, 8 is the cathode
           B. 8 is the anode, 7 is the cathode
           C. neither of the above is true

G182.   The circuit shown is:
           A. self-rectified
           B. half-wave rectified
           C. full-wave rectified
           D. none of the above

G183.   Transformers in an x-ray machine:
           A. work on the principle of electromagnetic induction
           B. have no moving parts
           C. require alternating current
           D. are housed in a tank of oil for electrical insulation and heat dissipation
           E. all of the above

G184-186. Match the device found in an x-ray circuit with its purpose.
A. allows current to flow in one direction only
B. increases or decreases voltage
C. thermionic emission
D. measure time of exposure
E. measures tube current

G184.   Transformer   B
G185.   Milliammeter   E
G186.   Rectifier   A

G187.   A transformer has a primary coil with N1 turns, and primary voltage V1. If the secondary coil has N2 turns, the voltage across the secondary coil is given by:
A. $(N2/N1)^2 \times V1$
B. $(N1/N2) \times V1$
C. $(N2/N1) \times V1$
D. $(N1/N2)^2 \times V1$
E. $(N1/N2)^{1/2} \times V1$

$\frac{N2}{N1} = \frac{V2}{V1}$

G188.   Step-up transformers are immersed in oil because:
A. oil is a poor heat conductor
B. oil is a good electrical insulator
C. oil prevents rusting of the transformer core
D. oil increases the efficiency of the transformer

G189.   Relative to the input, the output of a step-up transformer exhibits the following characteristics:
A. increased power
B. decreased voltage
C. increased current
D. all of the above
E. none of the above

↑Voltage, ↓ Current, ↓ Power

G190.   Which of the following does *not* improve the heat capacity of the x-ray tube?
A. rotating anode   ✓ ↑ focal area keeps, focal spor small
B. small target angle ↑ ratio of actual: effective focal spot
C. dual focus   enables focus sore to be kept small
D. thermionic emission   not related
E. all of the above

G191.   The rotating anode gives:
A. small effective focus and large heat loading capacity
B. better controlled exposure time
C. less soft radiation
D. mechanical rectification
E. all of the above

G192.   Two filaments are found in some x-ray tubes to:
         A. function as a spare if one burns out
         B. produce higher tube currents by using both filaments simultaneously
         C. double the number of heat units the anode can accept
         D. choose the smallest focal spot consistent with the kVp/mA setting.

G193.   The purpose of the x-ray tube filament found in an x-ray circuit is to:
         A. allow current to flow in one direction only
         B. increase or decrease voltage
         C. create thermionic emission *produce e⁻ from heated filament*
         D. measure time of exposure
         E. measure tube current

G194.   Heat is generated in all of the following processes, but is considered useful only in:
         A. the operation of a transformer
         B. the production of x-rays in a target   *} cancer production*
         C. the emission of electrons from a filament
         D. the treatment of malignancies by ionizing radiation— *negligible*

G195.   The effective energy of any x-ray beam:
         A. is proportional to the atomic number (Z) of the anode material
         B. is proportional to the mAs
         C. is not affected by the added filtration
         D. will affect image density but not subject contrast
         E. is always less than the kVp   *usually ½ to ⅓ of kVp*

         mA - independent of bremsstrahlung   *x-ray energy produced by bremsstrahlung, indep of Z or mAs*

G196.   Full-wave, compared with half wave rectification:
         A. requires two rectifiers instead of one
         B. reduces the voltage ripple
         C. delivers the same exposure in half the time
         D. increases the effective energy of the beam → *only kVp, filtration, etc do this.*
         E. reduces the heel effect

G197.   Compared to a single phase x-ray circuit, a three phase circuit has:
         A. less ripple
         B. higher average beam energy
         C. higher average dose rate
         D. all of the above
         E. none of the above

5. X-Ray Production

5.1.

only will see Tungsten char x-rays with higher energies than 69 keV!

G198-201. Tungsten has a K-shell binding energy of 69.5 keV. The following can cause a tungsten atom to emit a 57 keV K_α x-ray (answer A for true and B for false).

(particle or photon)

G198. A   An incoming 80 keV projectile electron.
G199. A   An incoming 75 keV gamma-ray.
G200. B   An incoming 66 keV x-ray.
G201. B   A 58 keV projectile electron.

learn more about Auger e's!

G202.   Consider an atom with a binding energy of the K-shell electron of 30 keV. The binding energy of the M-shell electron is 0.7 keV. An electron with a kinetic energy of 25.3 keV is ejected from the M-shell as an Auger electron following L to K transition. The binding energy of the L-shell electron is ____ keV.
A. 1.4
B. 4
C. 4.7
D. 15
E. 29.3

K 30
L (4)
M .7 + 25.3 = 26
        30 - 26 = 4

G203-212. Compare spectrums I and II in the diagram below. Choose the most appropriate answer from the list below for the next 10 questions.
A. spectrum I
B. spectrum II
C. both
D. neither
E. cannot be determined

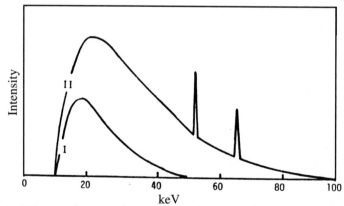

G203.   Characteristic radiations appear between 50 and 70 keV. B
G204.   Low energy photons have been removed by filtration.   C
G205.   Maximum photon energy is 100 keV. B
G206.   Minimum photon energy is 50 keV. D
G207. D The K characteristic x-rays have been removed by filtration. E

I doesn't have II was spike

G208.   Maximum kVp is 50 kVp. A
G209.   Higher exposure rate. B   *auc II > auc I*
G210.   Higher HVL. B   *HVL↑ as KVP↑*
G211.   Produced by three-phase or constant potential generator. E
G212. A Will produce relatively less scatter in soft tissue. ✗ *→ Scatter increases w/ KVP*

G213.   An atom has a K-shell binding energy of 70 keV, an L-shell energy of 8 keV and an M-shell
         energy of 1 keV. If an electron makes a transition from the L-shell to the K-shell, the emitted
         radiation will be _____ keV:
         A. 8   K 70
         B. 9   L 8
         C. 61
         (D.) 62   M 1
         E. 78

G214.   In a typical diagnostic x-ray beam from a tungsten target, characteristic radiation amounts
         to _____ percent of the total beam.
         (A.) 0-20
         B. 20-50
     ✓   C. 50-90
         D. 90-100

G215.   Molybdenum is usually used as a target for mammography tubes because it:
         A. has a higher melting point than tungsten
     ✗   B. has a higher efficiency for x-ray production   *the char x-rays are*
         (C.) produces characteristic x-rays of less than 20 keV *→ Ca⁺⁺ + soft tissue contrast*   *desirable*

G216.   100 keV electrons lose energy in an x-ray target via:
         (a) production of characteristic radiation, *.01%*
         (b) production of bremsstrahlung radiation, and *.04%*
     A   (c) heat loss. *99%*
         Rank these three phenomena in order of decreasing magnitude:
         A. abc   *most of energy lost as heat!*   *90% brem } approx*
         B. cab   *10% char } figures*
         C. bac
         D. bca
         (E.) cba

G217.   The maximum energy of characteristic radiation *always* increases with:
         A. energy of the incident electron *→ also true, but only*
         (B.) atomic number of the target   *if ...*
         C. filament current
         D. tube current   *see explanation*

G218-222. Consider an atom which has the following binding energies:
K-shell = 30 keV, L-shell = 4.0 keV, and M-shell = 0.7 keV
Match the energies of possible emissions for 50 keV incident electrons:
A. characteristic x-rays only
B. bremsstrahlung only
C. both A and B
D. neither A or B

G218.   B   3.7 keV
G219.   C   4 keV
G220.   B   25.3 keV
G221.   a   26 keV
G222.   B   49.3 keV

G223. The maximum photon energy in an x-ray beam is determined by the:
A. atomic number of the target
B. atomic number of the filter
C. kV across the tube
D. maximum mA
E. the kV wave form

5.2.

G224. It is decided to add 1 mm Al permanently to the photon beam filtration. This is done in order to reduce the:
A. load on the x-ray tube
B. scatter into the detection system
C. maximum field size
D. overall system latitude
E. patient skin dose

G225-228. Concerning x-ray beam output (answer A for true and B for false):

G225.   A   The quality of the beam depends on the wave form.

G226.   A   The quality of the beam is independent of the tube current (mA).

C227.   B   Small changes in filament current affect the beam quality.

G228.   A   A graph of intensity vs. photon energy is the best description of the x-ray tube output.

G229. The half-value layer (HVL) of an x-ray beam is:
A. equal to 1/2 the linear attenuation coefficient
B. directly proportional to the mass absorption coefficient
C. directly proportional to the linear attenuation coefficient
D. inversely proportional to the linear attenuation coefficient

G230. In general, the HVL does **not** depend on the:
A. peak kV
B. average kV
C. total filtration
D. radiation intensity
E. measuring geometry

G231. If a diagnostic x-ray beam has a first HVL of 3 mm aluminum, then adding 6 mm of Al to the filtration would decrease the intensity to:
A. 50%
B. somewhat more than 25%
D. somewhat less than 25%
C. 25%
E. 12.5%

G232. A narrow beam of monoenergetic photons is directed from a source 50 cm above the surface of a water-filled container 20 cm thick. The HVL is 10 cm of water for the photons. The exposure rate at the bottom of the container relative to that at the surface is about:
A. 50%
B. 25%
C. 17%
D. 13%
E. 6%

G233. The quality of a diagnostic or other low energy x-ray beam is usually determined by measuring:
A. the peak voltage across the tube
B. the effective kVp of the beam
C. the penetration through 10 cm of water
D. the amount of filtration in the beam
E. the half-value layer in aluminum

G234. The effective photon energy of an x-ray beam can be increased by:
A. increasing tube current
B. decreasing filtration
C. changing from half-wave to full-wave rectification
D. increasing the tube voltage
E. all of the above

G235.   A diagnostic x-ray beam has 2 mm Al filtration. If 1 mm of filtration were removed, the resulting beam would:
   A. have a higher dose rate and greater HVL
   B. have a lower dose rate and higher effective energy
   C. the same HVL but a higher dose rate
   D. have a HVL of 1 mm Al
   E. have a lower effective energy *& raise the dose rate*

↓ filtration
↑ dose rate
↓ HVL

6. Electromagnetic Radiation
6.1.

G236-240.   Match the effect in tissue and type of radiation with the name of the radiation (answers may be used more than once).
   A. ionizing elementary particles
   B. non-ionizing elementary particles
   C. ionizing photons
   D. non-ionizing photons
   E. none of the above

G236.   Beta-rays.   A

G237.   Heat radiation.   D

G238.   Visible light.   D

G239.   X-rays.   C

G240.   Ultrasound.   E   *mechanical wave thru medium*

G241.   Which of the following is *not* ionizing radiation?
   A. 10 keV x-rays
   B. ^{32}P beta-rays
   C. ^{226}Ra alphas
   D. 2 MHz ultrasound
   E. 6 MeV neutrons

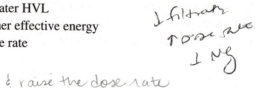

electromagnetic radiation
radio - micro - IR - US - UV - γ - X
↑λ   ↑ν
c = λν

G242-245.   Concerning the radiations listed on the right, which has:

G242.   D   The highest photon energy.   A. radio waves
                                                            B. visible light
G243.   A   The longest wavelength.   C. ultrasound
                                                            D. x-rays
G244.   D   The highest frequency.   E. all the same

G245.   C   The lowest velocity.

↳ speed of sound varies w/ medium

$E = \frac{hc}{\lambda}$
$E = hν = hc/λ$
$v < c$

G246.   Which of the following is in order of increasing wavelength?
         A. (radio waves) - (visible light) - (x-rays)
         B. (infrared red) - (radio waves) - (x-rays)
         C. (gamma-rays) - (ultraviolet) - (microwaves)
         D. (x-rays) - (visible light) - (ultraviolet)

6.2.

G247.   Which of the following properties is *not* true for electromagnetic radiation?
         A. zero mass and zero charge
         B. travels with velocity 3×10^8 m/s in free space
         C. includes radio waves
         D. is deflected by a magnetic field
         E. includes infrared

G248.   Which of the following is most descriptive of the difference between x-rays and gamma-rays?
         A. energy
         B. velocity
         C. field
         D. origin
         E. all of the above

G249.   Visible light has a wavelength of approximately 6×10^{-5} m. Cobalt-60 gamma-rays have a
         wavelength of 1×10^{-10} m and an energy of 1.2 MeV. What is the approximate energy of the
         light photons?
         A. 720 MeV
         B. 2 eV
         C. 2×10^{-6} eV
         D. 2×10^{-4} eV

G250.   The output of a fluoroscopic unit is 10 mR/min at 50 cm. The output at 75 cm will be
         _____ mR/min.
         A. 6.6
         B. 15.0
         C. 22.5
         D. 5.0
         E. 4.4

G251.   If a gamma source produces an exposure of 100 mR at 50 cm, the exposure at 100 cm will be:
         A. 400 mR
         B. 200 mR
         C. 100 mR
         D. 50 mR
         E. 25 mR

G252. The frequency of a photon of wavelength 6 cm is:
A. 18 Hz
B. 2×10^{-10} Hz
C. 5000 MHz
D. 5 MHz
E. 1.8×10^{-6} HZ

G253. Regarding electromagnetic radiation:
A. wavelength is directly proportional to frequency
B. velocity is directly proportional to frequency
C. energy is directly proportional to frequency
D. energy is directly proportional to wavelength
E. energy is inversely proportional to frequency

7. Photon Interactions with Matter

7.1.

G254. In the interaction of ionizing radiation and matter, absorption is associated with the transfer of energy:
A. to electrons
B. to photons
C. only in the photoelectric process
D. only from incident photons
E. only from charged particles

G255. The process whereby energy is transferred from a photon beam to electrons in the medium is called:
A. attenuation
B. absorption
C. bremsstrahlung
D. scatter

G256. Which of the following is **not** true? The linear attenuation coefficient:
A. is the fraction of incident radiation lost per unit thickness of absorbing material
B. equals the mass attenuation coefficient multiplied by the density of the material
C. usually has units of per cm
D. increases with increasing photon energy and decreasing atomic number

G257. The total mass absorption coefficient for photons:
A. has a minimum value for energies between 1 and 10 MeV
B. decreases as the atomic number of the absorber increases
C. is dominated by photoelectric effect at high energies
D. has a maximum value for energies between 1 and 10 MeV

G258. The quantity $e^{-\mu x}$ represents the:

$I = I_o e^{-\mu x}$

A. fraction of the primary photons attenuated by x cm of attenuator
B. fraction of the primary photons transmitted by x cm of attenuator ✓
C. average distance a primary photon will travel before it is attenuated
D. kinetic energy released by primary photons in x cm of attenuator

G259. A monoenergetic photon beam whose linear attenuation coefficient is 0.0693 cm^{-1} traverses 10 cm of a medium. The fraction of the beam that is transmitted is _____.

A. 0.01
B. 0.10
C. 0.37
D. 0.50
E. 0.69

$e^{-(.0693)(10)} = e^{-.693} = \frac{1}{2}$

G260. In the formula $I_x = I_o \exp(-\mu x)$, μ represents:

A. the thickness of the absorber
B. the initial beam intensity
C. the mass attenuation coefficient
D. the linear attenuation coefficient ✓
E. the half-value thickness

G261. The mass attenuation coefficient is similar for most materials (except those containing hydrogen) when:

A. the photoelectric effect predominates ~ $\frac{z^3}{E^3}$
B. pair production predominates ~ 2
C. only Compton interactions occur ✓ - depends on e⁻ density
D. photonuclear disintegration predominates
E. none of the above

G262. If μ is the linear attenuation coefficient of a sheet of material of thickness x, and I_o and I_x are the intensities of the incident and transmitted beams respectively, which of the following is true?

A. the HVL is the thickness, x, for which $I_x = I_o \exp(-0.693) = 50\%$
B. the HVL is the thickness, x, for which $I_x = I_o/2$ ✓
C. the HVL is $0.693/\mu$ ✓
D. all of the above
E. none of the above

G263. If the linear attenuation coefficient is 0.05 cm^{-1}, the HVL is:

A. 0.0347 cm
B. 0.05 cm
C. 0.693 cm
D. 1.386 cm
E. 13.86 cm

$HVL = \frac{.693}{.05} \quad \frac{.693}{\mu}$

G264.　　The fractional number of photons removed from a beam per cm of absorber is:
　　　　A. the linear-absorption coefficient
　　　　B. the mass-energy absorption coefficient
　　　　C. the scatter coefficient
　　　　D. the mean attenuation length μ = linear-absorption coeff × 1.44

G265.　　The fraction of photons absorbed after passing through n half-value layers of an absorber which has a linear absorption coefficient of μ will be:
　　　　A. $e^{-n\mu}$
　　　　B. $e^{+n\mu}$
　　　　C. $1 - e^{-n\mu}$
　　　　D. $e^{-(0.693n/\mu)}$
　　　　E. $1 - e^{-(0.693n)}$ ✓

$HVL = \frac{.693}{\mu}$

$I = I_0 e^{-\mu x}$

$abs = 1 - e^{-\mu x}$

$1 - e^{-\mu x}$

$\frac{\mu}{\mu} = .693$

$I = I_0 e$

$(e^{-.693} = .5)$

$(1.e.$

G266.　　CT or Hounsfield numbers are linearly related to:
　　　　A. mass density
　　　　B. electron density
　　　　C. linear attenuation coefficient
　　　　D. mass attenuation coefficient
　　　　E. effective atomic number

$CT\# = 1000 \left(\frac{\mu - \mu_{H2O}}{\mu_{H2O}} \right)$

G267.　　The photoelectric mass attenuation coefficient varies with:
　　　　A. $Z^1 E^1$
　　　　B. $Z^2 E^2$
　　　　C. $Z^3 E^3$
　　　　D. $Z^3 E^{-3}$ ✓
　　　　E. $Z^2 E^{-2}$

7.2.

G268-270. The following measurements are made in good geometry for a photon beam.

Added Filtration (mm Al)	Exposure (mR)
0	200
0.5	168
1.0	142
2.0	108
3.0	86
4.0	70
5.0	58
6.0	50
8.0	36

(handwritten: 75 (25%))

G268.   The second half-value layer is approximately_____ mm Al.

A. 2.0
B. 2.3
C. 2.9
D. 3.6
E. 6.0

(handwritten: Don't forget to subtract out the 1st HVL when calc the 2nd HVL)
(handwritten: 1st HVL ≈ 2.5 $\frac{6.5}{-2.5} / 3.5$)

G269.   If an additional 1 mm Al filtration is added to the beam, the first HVL will now be approximately _____ mm Al.

A. 2.0
B. 2.3
C. 2.9 ✓
D. 3.6
E. 6.0

(handwritten: 1mm → 142)
(handwritten: $\frac{142}{2}$ (= 1 HVL) → 70 4.0 mm Al $\frac{-1mm}{= 3}$)
(handwritten: 140 keV source)

G270. (B) The measurements could have been from a technetium-99m source (answer A for true and B for false).

(handwritten: Tc99m is a monoenergetic source! 140 keV)

G271.   Three tenth-value layers (TVL) will have approximately the same protective effect as ____ half-value layers (HVL).

A. 5
B. 10 ✓
C. 15
D. 20
E. 25

(handwritten: $(.1)^3 = .001$ $(.5)^x = .001$ x ≈ 10)
(handwritten: $.5^x = .001$)
(handwritten: 3 HVL ≈ 1 TVL)

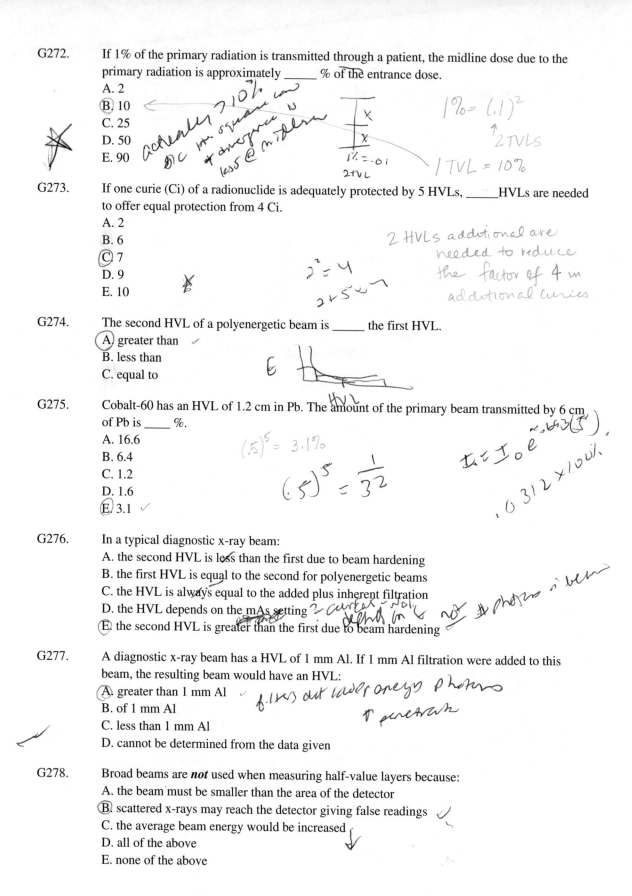

G272. If 1% of the primary radiation is transmitted through a patient, the midline dose due to the primary radiation is approximately _____ % of the entrance dose.
A. 2
B. 10
C. 25
D. 50
E. 90

$1\% = (.1)^2$
2 TVLs
1 TVL = 10%
$1\% = .01$
2 TVL

G273. If one curie (Ci) of a radionuclide is adequately protected by 5 HVLs, _____ HVLs are needed to offer equal protection from 4 Ci.
A. 2
B. 6
C. 7
D. 9
E. 10

$2^2 = 4$
2+5

2 HVLs additional are needed to reduce the factor of 4 in additional curies

G274. The second HVL of a polyenergetic beam is _____ the first HVL.
A. greater than
B. less than
C. equal to

G275. Cobalt-60 has an HVL of 1.2 cm in Pb. The amount of the primary beam transmitted by 6 cm of Pb is _____ %.
A. 16.6
B. 6.4
C. 1.2
D. 1.6
E. 3.1

$(.5)^5 = 3.1\%$
$(.5)^5 = \frac{1}{32}$
$I = I_0 e^{-.643(5)}$
$.0312 \times 100\%$

G276. In a typical diagnostic x-ray beam:
A. the second HVL is less than the first due to beam hardening
B. the first HVL is equal to the second for polyenergetic beams
C. the HVL is always equal to the added plus inherent filtration
D. the HVL depends on the mAs setting
E. the second HVL is greater than the first due to beam hardening

G277. A diagnostic x-ray beam has a HVL of 1 mm Al. If 1 mm Al filtration were added to this beam, the resulting beam would have an HVL:
A. greater than 1 mm Al
B. of 1 mm Al
C. less than 1 mm Al
D. cannot be determined from the data given

G278. Broad beams are **not** used when measuring half-value layers because:
A. the beam must be smaller than the area of the detector
B. scattered x-rays may reach the detector giving false readings
C. the average beam energy would be increased
D. all of the above
E. none of the above

G279.   Under what conditions will the second HVL of a beam of photons be approximately equal to the first HVL?
A. the initial beam is monoenergetic
B. the initial beam is polyenergetic
C. the beam is less than 100 kVp
D. this can never be true

7.3.

G280.   For the same exposure to a diagnostic beam, the highest absorbed dose will occur in:
A. fat
B. soft tissue
C. lung
D. bone
E. (all the same)

G281-285.   At x-ray energies between 40 and 100 keV, _____ absorbs *less* energy than _____ per gram (answer A for true and B for false).

G281.   Fat, muscle

G282.   Muscle, bone

G283.   Iodine, bone

G284.   Fat, air

G285.   Muscle, air

G286.   Absorption without scatter in soft tissue occurs in which type of interaction?
A. coherent
B. photoelectric
C. Compton
D. pair production

G287-291.   For the next six questions concerning the interaction of radiation and matter, answer A for true and B for false:

G287.   A photon may undergo a photoelectric interaction followed by a pair production interaction.

G288.   A photon may undergo two consecutive photoelectric interactions.

G289.   A photon may undergo a photoelectric interaction followed by a Compton interaction.

G290.   A photon may be totally absorbed by an atom.

G291.   The probability of a photoelectric interaction increases rapidly with energy.

G292. Two adjacent absorbers with the following physical properties are irradiated in a photon beam.

Absorber	A	B
Atomic number (Z)	7	14
Mass density (ρ)	1	2

The ratio of the photoelectric components of the linear attenuation coefficients (t_B/t_A) is approximately:
A. 2 to 1
B. 4 to 1
C. 8 to 1
D. 16 to 1
E. none of the above

G293-296. A 90 keV photon beam interacts with a soft tissue atom whose K-shell binding energy is 10 keV. An electron is emitted with a kinetic energy of 80 keV. (Answer A for true and B for false.)

G293. This is an example of Compton scattering.

G294. Characteristic radiation will be emitted.

G295. Auger electron emission is possible.

G296. The electron will be absorbed within 1 cm of its origin.

G297. The photoelectric component (τ/ρ) of the mass attenuation coefficient (μ/ρ) for water at 100 keV is about 0.004 cm^2/gm. The effective atomic number (Z) of water is about 8. The atomic number of lead is 82. The value of τ/ρ for lead at 100 keV is approximately _____ cm^2/gm.
A. 400
B. 40
C. 4
D. 0.4
E. 0.04

G298-302. In photoelectric interactions (answer A for true and B for false):

G298. K, L, and M characteristic x-rays may be produced if the photon energy is greater than the binding energy of the K-shell electrons.

G299. Photoelectric interactions occur only with loosely bound electrons.

G300. The probability of occurrence is greatest when the incident photon energy is a little less than the binding energy of the electron.

G301. The probability of the interaction is proportional to Z^3 per gram.

G302. In tissue, most of the released energy is locally absorbed.

G303. When a 10 keV photon undergoes a photoelectric interaction with a K-shell electron of binding energy 6 keV:

A. a 4 keV photoelectron is emitted *~*
B. a characteristic x-ray is emitted *✓*
C. the photon is ~~not scattered~~ with reduced energy
(D.) all of the above   *completely absorbed by e⁻*

G304. In water, photoelectric and Compton interactions are equally probable (for monoenergetic photons) at about:

A. 0.25 keV
B. 5.0 keV
(C.) 25 keV   *30*   *from 25 keV to 25 MeV*
D. 50 keV
E. 100 keV

G305. In diagnostic x-ray systems, filters are used to "harden" the beam. This process is mainly due to:

A. coherent scattering
(B.) photoelectric effect — *more likely to interact c low energies take out low E photons*
C. Compton effect
D. B and C
E. A, B and C

G306. On a relative scale, which of the following photoelectric interactions is most probable?

A. 30 keV x-ray and fat ($Z_{eff} = 6.3$)
B. 50 keV x-ray and bone ($Z_{eff} = 13.8$)
C. 70 keV x-ray and iodine ($Z_{eff} = 53$)
D. 70 keV and fat
(E.) 30 keV and iodine

7.4.

G307. A 5 keV photon undergoing classical scatter would be most likely to lose _____ % of its energy in the process.

(A.) zero   *no energy is lost*
B. 10
C. 33
D. 50
E. 90

G308-309. Match the following interactions and properties.
A. bremsstrahlung
B. photoelectric
C. Compton =scatter + absorption
D. pair production
E. coherent → scatter but no absorption

G308. E Attenuation without absorption.

G309. C Source of the most scattered photons in a diagnostic beam.

G310-315. Concerning the interaction of radiation and matter (answer A for true and B for false):

G310. A A photon may undergo a Compton interaction followed by a pair production interaction. AB

G311. A A photon may undergo two consecutive Compton interactions. A
after scatter photon produced

G312. A A photon may undergo a Compton interaction followed by a photoelectric interaction. A

But not PE b/c abs. process

G313. A The wavelength of a scattered photon may be greater than the wavelength of the incident photon. A
$E = h\nu = \frac{hc}{\lambda}$

G314. A A photon may scatter from an atom without losing energy. B *Coherent scattering!*

G315. B The probability of an 80 keV photon being scattered is greater in Sn (Z = 50) than in Al (Z = 13). A
only PE, no scattering, just an e⁻ released
depends on e⁻ density
↑ PE for Sn (more absorb) so Al more compton scatter

G316-319. In Compton interactions (answer A for true and B for false):

G316. B The photon changes direction but does not lose energy. B
$E_\gamma = E_{ce} + E_{cp}$

G317. B The electron may acquire any energy from zero up to the energy of the incident photon. A
$E_{max} = E_{h\nu} - E_{h\nu min} = h\nu(1 - \frac{1}{1+4E_{h\nu}})$

G318. A The maximum energy is imparted to the electron when the photon is scattered at 180°.

G319. B A neutrino is emitted.

G320. In a Compton interaction:
A. the photon is totally absorbed by the Compton electron
B. a characteristic x-ray and an electron are emitted
C. a Compton electron can be backscattered $0 \to 90°$
D. a photon of reduced energy can be backscattered
E. all of the above

no characteristic xrays from outer shell

G321.   If a technologist were to stand 2 m away from a patient during fluoroscopy (outside the primary beam) the dose received by the technologist would be mainly due to:
A. Compton electrons
B. photoelectrons — *low range*
C. Compton scattered photons
D. characteristic x-rays generated in the patient — *low range*
E. coherent scatter

G322.   The energy of backscattered photons is less than half the energy of the original photon for:
A. all x-rays
B. gamma-rays below 150 keV
C. x-rays in the diagnostic range
D. photons above 1 MeV
E. none of the above

as E↑, more % of E goes to e⁻ than γ

rules of thumb:
@ 100 keV, backscattered γ 70% of E incident
@ 1 MeV " " " 20% " "

G323-327.   For the next five questions, pick the most appropriate interaction process:
A. coherent scattering
B. photoelectric effect - *inner shell e⁻*
C. Compton scattering
D. pair production

Megavoltage → 250 keV + sideway cath 500 keV

G323.   c   Dominant at photon energies of 100 keV to 2 MeV in tissue. c

G324.   A   No energy is transferred (or locally absorbed). A

G325.   C   Probability (per unit mass) does not depend on the Z of the attenuating medium. c

G326.   C   Involves a free electron. *outer shell e⁻*

G327.   D   Increases with increasing photon energy. D

G328.   The ratio of Compton interactions in one gram of hydrogen to one gram of water is approximately:
A. 0.5
B. 1.0
C. 2.0
D. dependent on the photon energy
E. the ratio of the density of hydrogen to water

angle 1e⁻   all our atoms have per nucleon (p)   1 e⁻ per nucleon (n,p)
all elements are same (n,p)
so H has 2× e⁻ density

G329.   The binding energy of a K-shell electron in lead is 81 keV. After a Compton interaction involving 511 keV photons, backscattered electrons:
A. do not exist
B. have an energy of 255.5 keV
C. have an energy of 430 keV
D. have an energy of 511 keV
E. have a spectrum of energies from 0 to 511 keV

e⁻ can not be backscattered

G330.   A 10 MeV photon undergoes a Compton interaction. The backscattered photon has an energy of 255 keV. What angle does the Compton electron make relative to the direction of the initial photon?
A. cannot be determined from the information given
B. 0°
C. 90°
D. 180°
E. none of the above

G331.   Which of the following statements about Compton interaction is true?
A. Compton electrons can be ejected both forward and backward
B. Compton interactions have no effect on backscatter
C. a secondary photon scattered in the direction of the primary loses the most energy
D. the most energetic Compton electrons are those ejected at angles close to 90°
E. a secondary photon scattered at 180° cannot have an energy greater than 256 keV

7.5.

G332.   Which of the following results from pair production and eventually undergoes annihilation?

	Charge	Rest Mass
A.	+1	0.511 MeV
B.	+1	about 930 MeV
C.	0	0
D.	0	about 930 MeV
E.	-1	0.511 MeV

G333-337.   For the next five questions concerning pair production interactions, answer A for true and B for false:

G333.   The threshold energy for pair production is 0.511 MeV.

G334.   Electrons and positrons are produced.

G335.   The total kinetic energy of an incident photon is divided between the kinetic energies of an electron and a positron.

G336.   The annihilation of a positron produces 1.02 MeV photons.

G337.   The electron and positron are emitted in opposite directions.

G338.   The energy absorbed in pair production is:
A. the incident photon energy E_i
B. $E_i - 1.02$ MeV
C. $E_i -$ recoil energy of the nucleus

G339.   After a 2.044 MeV photon undergoes pair production, the following will always occur:
A. production of a pair of 511 keV photons *during annihilation*
B. production of a single 511 keV photon
C. production of a pair of 511 keV positrons
D. production of a pair of 1.022 MeV electrons

7.6.

G340-344. Match the numbered curves below with the appropriate mass attenuation coefficient:
A. Compton effect
B. pair production in lead
C. photoelectric effect in water
D. total absorption in water
E. total absorption in lead

G340.   Curve #1   E   *↑ ∝ Z³ for PE*
                          ↑ ∝ Z for PP

G341.   Curve #2   D

G342.   Curve #3   C

G343.   Curve #4   A

G344.   Curve #5   B

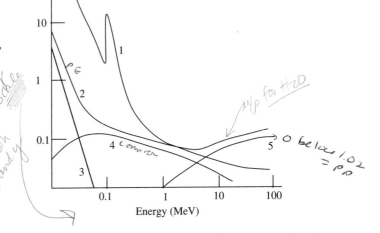

G345-350. Which interaction of an x-ray with matter is most likely to occur for the following situations?
(Answer A for photoelectric, B for Compton and C for pair production.)

G345.   A   A 35 keV x-ray incident on a vessel filled with iodine contrast material. *iodine K edge = 33.2 keV*
G346.   B   A 100 keV photon incident on muscle tissue.   B
G347.   B   A 15 MeV photon incident on muscle tissue.   C x
G348.   A   A 18 keV photon incident on fatty breast tissue.   A
G349.   B   A 90 keV photon incident on bone.   B *(40 keV = break point)*
G350.   A   A 90 keV photon incident on a lead barrier.   B x *Pb K edge = 88 keV*

most interactions occur b/w 25 keV and 25 MeV
Compton interactions

G351.   The mass attenuation coefficient for photons in water:
A. decreases continuously with energy below 25 MeV x
B. decreases to about 3 MeV, then rises again *for lead*
C. increases continuously with energy below 25 MeV x
D. rises to a peak at about 3 eV x
E. none of the above

G352-354. Match the curve in the diagram below with the interaction it represents.

A. photoelectric
B. coherent scatter
C. photonuclear disintegration
D. Compton
E. pair production

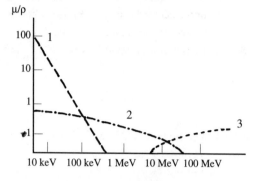

G352. A Curve #1   A

G353. D Curve #2   D

G354. E Curve #3   E

G355.   In the diagnostic energy range, when the atomic number of an absorber changes from Z = 20 to Z = 40 the mass attenuation coefficient:
A. increases ✓
B. decreases   pe
C. stays the same

G356-359. The diagram below applies to the next four questions.

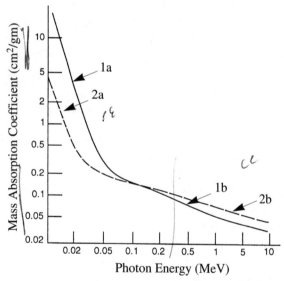

G356.   Curves la and 2a represent x-ray and gamma-ray interactions due to:
A. coherent scattering
B. photoelectric effect ✓
C. Compton scattering
D. pair production
E. photodisintegration

G357.   Curves lb and 2b represent:
A. coherent scattering
B. photoelectric effect
C. Compton scattering ✓
D. pair production
E. photodisintegration

G358.   Comparing the atomic number of the two tissues:
A. 1 and 2 are approximately equal
B. 1 is greater than 2 ✓
C. 2 is greater than 1
D. cannot tell from this plot

G359.   Comparing the densities of the two tissues:
A. 1 and 2 are approximately equal
B. 1 is greater than 2
C. 2 is greater than 1
D. cannot tell from this plot ✓ → b/c of the y-axis

mass abs coef not related to density

had they been plotted as a linear µ
you could tell

8. Charged Particle Interactions

but linear att coef α density

mass abs coef = $\dfrac{\text{linear att}}{\mu/\rho}$

G360.   The radiation which has the greatest range in tissue is:
A. an 8 MeV alpha particle   *micron*
B. a 2 MeV beta particle ✓   *1 cm*   *½ = 1 cm*
C. a 10 keV Auger electron .
D. a 10 keV proton –
E. a 1 MeV positron   *1/? = .5 cm*

G361.   An electron, a proton, and an alpha particle each have 20 MeV kinetic energy. Which of the following statements is true?
A. the alpha particle travels at almost the speed of light *slow*
B. the alpha particle has the least total energy → *has the most total energy* *b/c of its high rest mass.*   *2p + 2n*
C. the proton has the highest total energy
D. the electron travels almost at the speed of light ✓ *(@ 1 MeV)*
E. none of the above   *low rest mass – high ke*

G362.   Electrons lose energy when passing through matter by:
A. 1 & 2 only
B. 3 & 4 only ✓
C. 1, 3, & 4 only ✓
D. all are correct

1. production of bremsstrahlung ✓
2. photoelectric interactions *photons*
3. collision with other electrons ✓
4. production of delta rays ✓ ✓

9. Film Characteristics and Image Quality

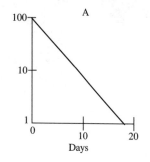

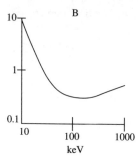

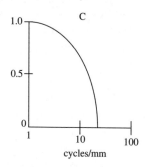

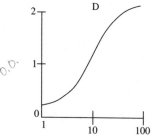

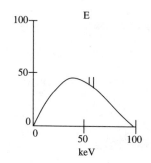

G363.   Which of the above graphs represents the shape of an H & D curve?   D

OD vs. Dose

G364.   Two x-ray films, each with optical density of 1.5, are placed on top of one another. The fraction of incident light transmitted through the "sandwich" is:

A. 0.03
B. 0.015
C. 0.001
D. 0.0225

O.D. is additive   _O.D. = log I₀/I_   _I = I₀ × .001_

3 = log I₀/I   _I₀/I = 1000_

G365.   Two identical unexposed x-ray films are irradiated to different doses with an identical x-ray beam. After development, one film has a optical density (OD) of 2, and the other an OD of 1. The relative doses received by the two films are:

A. 2:1 – if OD proportional to linear dose
B. 10:1
C. 1.41:1 – if proportional to log dose
D. cannot be determined from given information – must know H+D curve

G366.   A grid improves the quality of a diagnostic x-ray primarily by:

A. attenuating primary photons
B. attenuating Compton scattered photons ✓
C. attenuating electrons produced by Compton scatter
D. attenuating electrons produced by photoelectric effect
E. attenuating coherently scattered photons

G367. Using a grid when taking a diagnostic x-ray does all of the following *except*:
A. increase the contrast of the image
B. increase patient dose for the same optical density on film
C. attenuate Compton scattered photons
D. absorb electrons produced by the photoelectric effect

G368. In a diagnostic radiograph the process mostly responsible for differential attenuation is:
A. coherent scatter
B. Compton interaction
C. photoelectric interaction
D. pair production

G369. Which of the following does *not* reduce patient dose (for the same optical density on the film)?
A. the use of screens
B. using a high kVp
C. the use of a high ratio grid
D. all of the above, since none reduce patient dose

10. Statistics

G370. A radioactive sample is counted for a ten minute interval many times, yielding a mean count rate of 1000 cpm. The most probable distribution is:
A. 68% of the measurements fall between 990 and 1010 cpm
B. 68% of the measurements fall between 936 and 1064 cpm
C. 68% of the measurements fall between 900 and 1100 cpm
D. 95% of the measurements fall between 936 and 1064 cpm
E. 95% of the measurements fall between 800 and 1064 cpm

G371. To achieve an estimated deviation of 2%, _____ counts must be collected.
A. 400
B. 1414
C. 2500
D. 10000
E. 40000

G372. A measurement is made in which 2500 counts are collected. There is a 96% probability that repeated measurements will yield between _____ and _____ counts.
A. 2300, 2500
B. 2400, 2500
C. 2400, 2600
D. 2450, 2550
E. 2500, 2700

Raphex Compilation ▲ General Questions ▲ Page 53

G373-378. Concerning the poisson distribution (answer A for true and B for false):

G373. B It is another name for the normal distribution.

G374. A It is due to random variations.

G375. B Photon distribution on an x-ray film is a poisson distribution.

G376. A Radioactive decay as a function of time is a poisson distribution.

G377. A The standard deviation (σ) increases as the number of measurements increases.

G378. B The percent standard deviation (%σ) increases as the number of measurements increases.

G379. D A radioactive sample is counted for 1 minute and produces 900 counts. The background is counted for 10 minutes and produces 100 counts. The net count rate and net standard deviation are about _____ counts.

    A. 800 + 28
    B. 800 + 30
    C. 890 + 28
    D. 890 + 30
    E. 899 + 30

11. Computers

G380.   The mainframe of a digital computer contains:
    1. core memory
    2. central processing unit (CPU)
    3. only random access memory
    4. software
    5. connectors for peripheral devices

    A. 1, 2, 5
    B. 1, 2
    C. 2, 4
    D. 1, 4
    E. 1, 2, 3, 4, 5

G381.   ROM is memory that can be:
    A. used and changed freely
    B. freely read but not written to
    C. repeatedly used to store output from an input device
    D. used only with alpha-numeric characters
    E. randomly accessed

G382.   A byte is a:
         A. device physically placed between a terminal and memory
         B. signal from a digitizer sent to memory
         C. control key on a terminal
         D. fixed number of bits   ✓ usually 8 or 16, which stores 1 unit of information

G383.   Information will be destroyed in/on a ____ when the computer power is turned off.
         A. floppy disk
         B. hard disk
         C. magnetic tape
         D. RAM   ✓
         E. ROM

G384.   List the following in terms of information storage capacity (highest to lowest):
         F = Floppy disk   lowest   1 byte = 8 bits   usually
         H = Single hard disk
         M = Multiplatter hard disk
         O = Optical disk   highest
         T = Magnetic tape reel (1600 BPI, 2400 ft.)

         A. HFOMT
         B. TFMOH
         C. MTFHO
         D. OMTHF   ✓
         E. FHMTO

G385.   A 16-bit word computer can directly address a maximum of how many different locations?
         A. 16
         B. 32,581
         C. 58,325
         D. 65,536 2^{16}
         E. 130,036

G386-390. Indicate whether each of the following computer components would be designated as an
         input/output device or a storage device.
         A. input/output
         B. storage

G386. A Typewriter terminal

G387. B Hard magnetic disk

G388. B Magnetic tape

G389. A CRT terminal

G390. B Optical disk

G391-400. Concerning digital computers (answer A for true and B for false):

G391. B ROM stands for random order memory. Memory in which data can be placed in any desired order. *read only*

G392. A RAM stands for random access memory. Memory in which groups of storage locations can be addressed directly and independently of each other. A

G393. B Real time: information is gathered and processed at a precise rate of one image per second rather than at other random rates. B *@10 30/sec* *key events are processed in (imperceptible lapse time) — integration time else clock time ≈ 1/30 sec*

G394. A Word: a set of consecutive bits treated as an entity within the computer and occupying one storage location in memory.

G395. B Byte: a binary digit. The smallest unit in the binary system used to represent 1 or 0. *BIT* *Byte = adjacent binary digits* *8 16*

G396. A File: a collection of interrelated records treated as a unit.

G397. A Microprocessor: a single large scale integrated circuit (chip) that has capabilities of performing arithmetic and logical operations on bytes and words.

G398. A Modem: a device that converts a digital signal into a frequency coded signal for transmission on a carrier wave over a communications line such as a telephone line.

G399. Arrange in order of increasing size:
A. block-byte-word-bit
B. bit-byte-word-block ✓
C. byte-bit-block-word
D. word-byte-bit-block

a block is 512 bytes

G400. A computer system can be configured so that the boot instructions are read from any of the following *except*:
A. random access memory *RAM → could be lost when electrical power is turned off*
B. read only memory
C. a floppy disk ✓
D. a hard disk

G401. If all eight bits in a byte are set to one, the decimal value of the byte is:
A. 1
B. 2
C. 8 *$2^8 = 255$?*
D. 255
E. 256

G402.   List the following in typical access time (slowest to fastest).
         F = floppy disk
         H = hard disk   *(fastest)*
         T = magnetic tape *(slowest)*

         A. HFT
         B. FTH ✓
         C. TFH
         D. FHT
         E. THF

                                    → *dividing a task into pieces*

G403.   Parallel processing refers to:
         A. running multiple tasks simultaneously *on different processors*
         B. using multiple processors to increase speed *→ both are r.f.*
         C. computer networking
         D. sharing peripheral devices between computers

12. Protection & Dose Measurement

NCRP reports

12.1.

G404.   The average natural background is made up of cosmic radiation, terrestrial radiation and:
         A. fallout
         B. scattered medical radiation
         C. nuclear plant releases   *avg. US. natural background*
         D. radioactive waste disposal contamination   *annual radiation ~*
         E. internal radiation $(\sim 40\ mrem\ from\ ^{40}K)$ *100 mrem*

G405.   The most significant source of man-made radiation dose to the population as a whole is from:
         A. high altitude air travel
         B. television receivers and other consumer products
         C. fallout from nuclear weapons exploded in the atmosphere
         D. diagnostic radiological examinations *② nuclear medicine exams*
         E. nuclear reactor effluents

G406.   Studies of the affects of the atomic bombs dropped on Japan during World War II indicate that
         the probability of inducing cancer in a large population that is irradiated with 1 rem is
         about _____ during that population's lifetime.   *from BEIR III study*
         A. 1 in 10
         B. 1 in 1000
         C. 1 in 10,000
         D. 1 in 100,000
         E. 1 in 1,000,000

G407.　According to recent estimates, the largest contribution to the radiation exposure of the United States population as a whole is from:
A. medical x-rays
B. nuclear medicine procedures
C. radon in the home
D. the nuclear power industry
E. nuclear weapons production and testing

G408.　The principal hazard from indoor radon involves:
A. whole body doses from gamma-rays
B. skin doses from beta-rays
C. lung doses from alpha emission　*products are inhaled*
D. bone doses from deposited radionuclides

G409.　According to NCRP Report #91, the quality factor for high LET radiations such as neutrons should generally be:
A. 1
B. 2
C. 5
D. 10
E. 20

every thing else ...)
$\alpha \cong 10$

G410.　Radon represents an environmental hazard primarily in:
A. high rise apartment buildings
B. wooden buildings
C. basements and ground floor apartments
D. outdoors

12.2.

G411.　According to NCRP Report #116, the recommended maximum annual dose equivalents for radiation workers' (a) whole body, (b) an eye lens, and (c) any single organ in mSv, are:

	(a)	(b)	(c)
A.	5	15	50
B.	5	5	5
C	10	50	100
D.	50	10	50
E.	50	150	500

100 rem = 1 Sv
aim ≈ 10 mSv
Whole body 5000 mrem 5 rem = 50 mSv
eye 15 rem = 150 mSv
single organ 50 rem

G412.　10 CFR Part 20 recommends that radiation monitoring devices shall be used by personnel who:
A. require radiation for medical purposes
B. work in an area that contains a radiation source
C. could be exposed to more than 1/10 of the maximum permissible dose (MPD)
D. could be exposed to more than 2 mR in any one hour
E. all of the above

G413-417. Filters are used in film badges to (answer A for true and B for false):

G413. B   Convert x- and/or gamma-ray energy to visible light to expose the film. B

G414. B   Shield a portion of the film for a base density reading. A

G415. A   Discriminate between different types of radiation. A   *open window = β ray*
crude

G416. A   Discriminate between different radiation energies. A

G417. B   Reduce the exposure to the user. B

G418.   Film badges:
    A. can measure only the total dose of radiation, but cannot distinguish between low and high energy x-rays
    B. can measure exposures of 1 mR
    C. are insensitive to heat
    D. can be used to determine dose received from the optical density of the film
    E. none of the above

(film badges cannot measure <20 mR accurately)

12.3.

G419.   The exposure rate constant of a radionuclide is 12.9 R-cm^2 /mCi-hr. How many HVLs are required to reduce the exposure rate at 1 meter from a 10 mCi source to 2 mR/hr?
    A. 1
    B. 2
    C. 3
    D. 6

$$\Gamma A \left(\frac{1}{d^2}\right)$$

$$12.9 \frac{R\,cm^2}{mCi\,hr} \cdot 10\,mCi \cdot \left(\frac{1}{100\,cm}\right)^2 = \frac{129}{10000} \frac{R}{m} = .0129\,R/m$$

$$12.9 \frac{R\,cm^2}{mCi\,hr}(10)\left(\frac{1}{100}\right)^2 = .0129 \frac{R}{hr}$$

3 HVLS → 12.9 mR

G420-424. A shielding design for a diagnostic or therapy installation, when there is no restriction on the beam direction, must (answer A for true and B for false):

G420. A   Consider all walls as primary barriers.
G421. B   Assign all walls a use factor (U) of 1.
G422. B   Assign any area which people may frequent an occupancy factor (T) of 1.
G423. B   Shield all areas to a radiation level of 0.1 rem per week.
G424. A   Shield such that unrestricted environments will **not** receive greater than 2 mR in any one hour.

$$B = \frac{P d^2}{W U T}$$

only walls that man ptbar ... *not possible, most walls are 1/4 or 1/16* ... *1/4 office console* ... *only for controlled areas* ... *noncontrolled areas are .002 rem/wk*

G425.   The occupancy factor (T) is changed from 1/16 to 1/2 and the activity (A) is doubled for a radiation source whose HVL is 0.3 mm Pb. In order to maintain the same level of protection, _____ mm Pb must be added.
    A. 0.3
    B. 0.6
    C. 0.9
    D. 1.2
    E. 1.5

T increases by 8   A increases by 2   a factor of 16 increase

$$2^x = 16$$
$$2^4 = 16$$
$$.3\,mm = 1\,HVL$$
$$.3 \times 4 = 1.2$$

G426.	The use of a diagnostic room is changed. It is estimated that the workload will double. The maximum beam energy and the use factor are the same. The change in shielding required at the console is:
   A. no change; shielding is adequate
   B. add 1 HVL
   C. add 2 HVLs
   D. remove 1 HVL
   E. insufficient information given

[handwritten annotations: workload, use factor; $\frac{WUT}{d^2}$ occupancy factor (=1 in this case); $B = \frac{P d^2}{WUT}$; $2^x = 2$, $x = 1$ HVL]

G427.	When calculating radiation barrier thickness requirements, the use factor, U, refers to:
   A. the weekly dose delivered at one meter from the radiation source
   B. the fraction of the operating time during which the area on the other side of the barrier is occupied
   C. the fraction of the operating time during which the radiation under consideration is directed toward the particular barrier
   D. the fraction of the work week during which an individual is in the area of interest
   E. the fraction of the week that the machine is in operation

G428.	In barrier calculations, the occupancy factor, T, refers to:
   A. the weekly dose delivered at one meter from the radiation source
   B. the fraction of the operating time during which the area on the other side of the barrier is occupied
   C. the fraction of the operating time during which the radiation under consideration is directed toward the particular barrier
   D. the fraction of the work week during which a radiation worker is in the area of interest
   E. the fraction of the week that the machine is in operation

G429.	The thickness of the shielding wall for an x-ray unit increases with all of the following *except*:
   A. occupancy factor of adjacent area
   B. use factor
   C. room size
   D. beam energy

G430.	A new storage area for nuclear medicine has been installed, and a measurement made in an adjacent clerical office indicates that the workers in that office would receive an annual dose of 1 rem. A prudent course of action would be:
   A. do nothing. The dose rate measured is within legal limits, even for non-radiation workers
   B. the dose rate measured is legal only for radiation workers and the workers should be issued film badges
   C. double the amount of shielding in the walls
   D. add one tenth-value layer of shielding to the walls
   E. only permit occupancy for 1/2 of each day

[handwritten annotation: max permissable dose to nonradiation workers = .1 $\frac{rem}{yr}$ for continuous exposure]

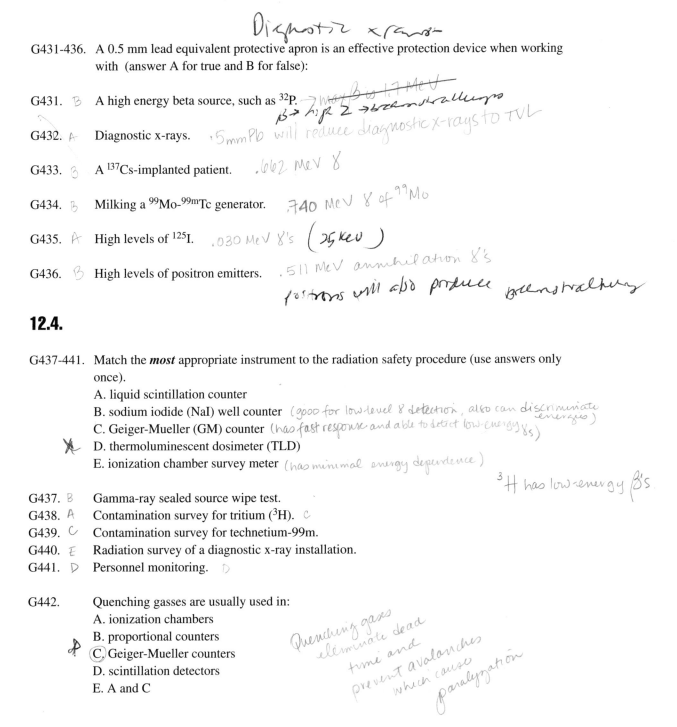

Diagnostic x-rays (handwritten title)

G431-436. A 0.5 mm lead equivalent protective apron is an effective protection device when working with (answer A for true and B for false):

G431. **B** A high energy beta source, such as ^{32}P. *→ may β is 1.7 MeV → bremsstrahlung*
β → high Z → bremsstrahlung

G432. **A** Diagnostic x-rays. *.5mm Pb will reduce diagnostic x-rays to TVL*

G433. **B** A ^{137}Cs-implanted patient. *.662 MeV γ*

G434. **B** Milking a ^{99}Mo-^{99m}Tc generator. *.740 MeV γ of ^{99}Mo*

G435. **A** High levels of ^{125}I. *.030 MeV γ's (25 keV)*

G436. **B** High levels of positron emitters. *.511 MeV annihilation γ's*
positrons will also produce bremsstrahlung

12.4.

G437-441. Match the ***most*** appropriate instrument to the radiation safety procedure (use answers only once).
A. liquid scintillation counter
B. sodium iodide (NaI) well counter *(good for low-level γ detection, also can discriminate energies)*
C. Geiger-Mueller (GM) counter *(has fast response and able to detect low-energy γ's)*
D. thermoluminescent dosimeter (TLD)
E. ionization chamber survey meter *(has minimal energy dependence)*

^{3}H has low-energy β's

G437. **B** Gamma-ray sealed source wipe test.
G438. **A** Contamination survey for tritium (^{3}H). *c*
G439. **C** Contamination survey for technetium-99m.
G440. **E** Radiation survey of a diagnostic x-ray installation.
G441. **D** Personnel monitoring. *D*

G442. Quenching gasses are usually used in:
A. ionization chambers
B. proportional counters
(C.) Geiger-Mueller counters
D. scintillation detectors
E. A and C

Quenching gasses eliminate dead time and prevent avalanches which cause paralyzation

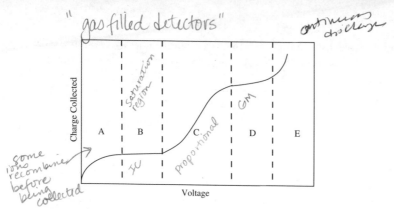

"gas filled detectors"   continuous discharge

Charge Collected

A | B | saturation region | C proportional | GM | D | E

some ions recombine before being collected

XU   proportional

Voltage

G443-447. The above graph shows the response curve for a gas filled detector. Give the appropriate region for each of the following descriptions:

G443. E   The region where current is passed in the absence of radiation.
            Conduction in gases happens here

G444. C   The pulse heights are proportional to the incident particle energy.

G445. B   An ionization chamber is operated in this region.

G446. D   A Geiger survey meter is operated in this region. (any event will trigger an avalanche → giving pulses of equal size)

G447. B   The saturation region where all charges are collected without multiplication. B

G448.   In general, which of the following detectors has the greatest energy dependence for x- and gamma-rays?
           A. thin window Geiger tube   relatively energy independent
           B. air wall ionization chamber
           C. LiF thermoluminescence dosimeter
           D. sodium iodide scintillation detector   Iodide has high Z, thus a ↑ PE absorption

G449.   Which of the following instruments is the most sensitive for locating radioactive sources?
           A. Geiger tube survey meter   → cascade effect amplifies a weak signal
           B. ionization type survey meter   requires only 1 ionization
           C. thermoluminescent dosimeter
           D. pocket dosimeter
           E. film-badge dosimeter

G450.   The quantity that an ionization chamber actually measures is:
           A. roentgen
           B. gray
           C. charge
           D. kerma
           E. voltage

G451. A Farmer thimble ionization chamber is calibrated at 22°C and 760 mm Hg pressure. What is the temperature-pressure correction factor if a measurement is made at a temperature of 18° and 750 mm Hg pressure?

$$\left(\frac{760}{750}\right)\left(\frac{273.2+18}{295.2}\right)_{(T_r)} = .999 \approx 1$$

(2) $273+22$

A. 0.81
B. 0.98
C. 1.00
D. 1.02

G452. Electron equilibrium is said to exist in a small mass when:

usually d_max depth where Kerma = Dose

A. the number of x-rays entering the mass is equal to the number of x-rays leaving it
B. the build-up of dose begins to slow down
C. absorption of the radiation becomes exponential
D. the number of electrons entering the mass is equal to the number of electrons leaving it ✓

G453. An instrument used for absolute measurements of exposure is:
A. Baldwin-Farmer ionization chamber
B. thermoluminescent dosimeter
C. free-air chamber → *only one that doesn't need to be compared to standard* b/c *1 Roentgen = 2.58×10⁻⁴ C/kg*
D. survey meter

G454. Geiger counters:
A. are sensitive enough to detect individual photons or particles ✓
B. are less sensitive than ionization chambers
C. are used for accurate measurements of exposure
D. operate on the principle of scintillation

G455. Calorimetry is a technique used for:
A. routine exposure measurements
B. routine dose measurements
C. absolute exposure measurements
D. absolute dose measurements → *but difficult to practice!*

G456. Ion-recombination can be a problem when using a:
A. calorimeter
B. Geiger counter
C. ionization chamber ✓ *voltage must be high enough*
D. TLD

12.5.

G457. The basic consideration in disposing of radioactive materials in the sanitary sewer system is:
A. contamination of the sewage system
B. risk to swimmers and bathers
C. fish death
D. entrance into the food and fresh water chain
E. evaporation into the air

G458.   A patient who had an iodine-125 seed prostate implant 3 years ago is admitted for a transurethral resection of the prostate (TURP). The activity implanted was 25 mCi. The half-life of iodine-125 is 60 days. Which of the following is true?
A. radiographs should not be attempted, as radiation from the seeds would fog the film
B. the physician performing the TURP should wear a Pb apron and Pb gloves
C. the TURP should not be attempted, because of the radiation hazard to OR staff
D. none of the above are true

G459.   Which of the following is true for low-level radioactive wastes, such as tubing and swabs contaminated with ^{99m}Tc ?
A. they can never be thrown away, since some activity always remains
B. they can be thrown away immediately, since the amount of activity is generally harmless
C. they can only be disposed of by a commercial rad-waste service
D. they can be stored until reaching background levels, then disposed of with other medical trash

12.6.

G460.   A "controlled area" is defined as:
A. any area around a radiation facility where the exposure rate is above background
B. an area which one cannot enter unless wearing a film badge
C. an area where the workers will not receive more than 10 mrem/week
D. an area where the exposure of workers is under the supervision of a Radiation Safety Officer

G461.   A *High Radiation Area* sign must be posted at the entrance of all areas where the exposure rate can be:
A. >100 mR/hr
B. >100 mR/hr but < 1000 mR/hr
C. >10 mR/hr
D. >100 mR/wk
E. none of the above

Answers

1.

G1.   D   Absorbed dose per roentgen in tissue requires knowledge of the f-factor.

G2.   E   The quality factor (Q) is an approximation for RBE.

G3.   D   Exposure (X) is defined as ionization in air per unit mass produced by x- or γ-rays. Dose (D) is defined as energy absorbed per unit mass by ionizing radiation. Exposure is limited to ionizing photons. Ionization produced directly by high energy particles is excluded. Dose is limited only to ionizing radiation. It may be produced by photons, energetic charged or uncharged particles. The rad and the gray (Gy) are both units of dose. 1 rad = 10^{-2} joules/kg. 1 Gy = 1 joule/kg. 1 Gy = 100 rad.

G4.   E   The energy absorbed by a mass of air from x- or γ-rays per roentgen is a constant, 0.87 cGy/R. It can be calculated by multiplying the number of ion-pairs produced per R per kg by the average energy necessary to produce one ion pair (w) and converting the product from eV to joules. w = 33 eV/ion pair.

G5.   D

G6.   C

G7.   D

G8.   A

G9.   C

G10.   A

G11.   D

G12.   B

G13.   E

G14.   A

G15.   E

G16.   C

G17.   C   A is absorbed dose (gray), and D is dose equivalent (sievert). Exposure is defined only for photons below 3 MeV; in the SI system units of coulomb/kg replace the roentgen.

G18.   C

G19.   C

G20.   E

G21.   C   All x-rays, gamma and betas used for diagnostic and therapy purposes have Q = 1. For neutrons the RBE varies with energy, but for protection a "worst case" value of 20 is used. See NCRP Report #91.

G22.   E   The SI unit of dose equivalent is the sievert. 1 sievert is 100 rem.

G23.   D   Q depends on the LET (linear energy transfer) of the radiation. Values used for radiation protection are:

Radiation type	Q
x-rays, electrons and gamma-rays	1
neutrons	20

G24.   D   Hertz (Hz) or cycles/sec is the unit of frequency. The "wave equation," where c is the speed of light, is c = wavelength × frequency.

G25.   B   Exposure is defined only for photon beams, and its SI unit is C/kg. At diagnostic energies, the absorbed dose to bone is greater than that to muscle for the same exposure, because of the predominance of the photoelectric effect.

G26.   D   "A" was the way the roentgen was originally defined; "B" is the definition in SI units; "C" is the roentgen to cGy conversion factor (f-factor) for air.

G27.   A   Absorbed dose is defined as the energy absorbed per gram, although in this case the integral doses would be equal.

G28.   C   The kerma is defined as the sum of the initial kinetic energies of all the charged particles liberated by indirectly ionizing radiation in a volume element. However, all of this energy may not be deposited within the volume element if the particle range is large compared to the size of the element, because some fraction of energy can be radiated as bremsstrahlung. Therefore, kerma can differ from the absorbed dose at the point of interest.

G29.   B

G30.   D

G31.   C

G32.   E

G33.   A   The modern definition of 2.58×10^{-4} C/kg of air (E) is obtained by changing the units from the original definition.

G34.   B   Amps = coulombs/second.

G35.   D   1 rad = 100 erg/gm. The SI unit for dose is the gray = joule/kg.

2.1

G36.   B   Proton, (+1, 930 MeV): highest combination of charge and mass.

G37.   A   Positron, (+1, 0.511 MeV); positrons are unstable.

G38.   A   Unlike charged particles attract.

G39.   C   Like charged particles repel.

G40.   A   Unlike charged particles attract.

G41.   A   Unlike charged particles attract.

G42.   B   There is no electric force between a charged and a neutral particle.

G43.   A   This particle has negligible mass; A is the only choice.

G44.   E   This particle has a charge of 2; E is the only choice.

G45.   C   C is the only choice with a charge of +1.

G46.   B   A, B and D have no charge; however, if C is a positron, the neutron must have a greater mass, so B is the correct choice.

G47.   C

G48.   B

G49.   C   In beta minus decay, a neutron changes into a proton, with the emission of a beta minus particle (electron) and an antineutrino. The latter share the available energy.

G50.   D   Pair production occurs in the field of the nucleus when a photon of energy at least 1.022 MeV disappears, and an electron and positron are created.

G51.   A   When a neutron is absorbed by the nucleus of a ^{235}U atom, fission occurs. Several neutrons are emitted, which can cause fission in adjacent ^{235}U nuclei, causing a chain reaction. Some of the fission products in reactor fuel rods are used in medicine (e.g., ^{137}Cs).

G52.   C   Neutrons are not charged particles, and generally interact with matter by transferring their energy to protons or other light nuclei, which then produce dense ionization tracts.

G53.   B   Alpha particles consist of 2 protons, each +1, and 2 neutrons, charge 0.

G54.   D   The neutron has no charge.

G55.   E

G56.   C

G57.   D   A photon carries energy but no charge.

G58.   E

G59.   B   Although most odd-numbered nuclei have a magnetic moment, only hydrogen nuclei, consisting of a single proton, are detected in most MR imaging.

G60.   D   Neutrinos have no charge and almost no mass, and rarely interact with matter.

G61.   A   Beta-rays emitted in negative beta decay are electrons.

G62.   E   The exact value = 939.55 MeV/0.511 MeV = 1839.

G63.   A   If the kinetic energy of a particle is >> than the rest mass, then the velocity is very close to c. Electrons have a rest mass of 0.511 MeV. Photons always travel at velocity c.

G64.   B

2.2

G65.   C   Most elements can have several stable and radioactive forms; an element (and its isotopes) can occupy only one row and column of the periodic table, governed by Z.

G66.   E   Isotopes of an element have the same Z (no. of protons or electrons), but different numbers of neutrons, and hence different A (mass no.). Chemically, all isotopes of the same element are identical.

G67.   D   Isotopes are forms of the same element, and thus have the same atomic number, Z, the number of protons, but different numbers of neutrons, thus different A (neutrons plus protons). Isobars have the same A but different Z. Isomers have the same A and Z, but different energy states. Isotones have the same number of neutrons but different Z.

G68.   C   Isotones have the same number of neutrons, but different numbers of protons, and hence different masses. Isobars have equal numbers of nucleons, and hence (nearly) identical masses. An isomer is a metastable state of a nucleus.

G69.   B   Isobars have equal numbers of nucleons, and nearly identical masses. The daughters are isobars, with atomic numbers differing by 2. Very few nuclei have this unusual decay pattern, ^{40}K and ^{64}Cu being among them.

2.3

G70.   C   The mass number (A) is defined as the number of nucleons (protons and neutrons) in the atomic nucleus.

G71.   C   The number of neutrons in a nucleus is the mass number (total no. of nucleons) minus the atomic number (no. of protons), i.e., $60 - 27 = 33$.

G72.   B   All isotopes of hydrogen have one proton and one electron. The mass number (A) is 3, so there are 2 neutrons.

G73.   C   The number of electrons is normally equal to the number of protons in the nucleus, which is the atomic number, Z.

G74.   D   1 amu = 931.5 MeV, or approximately 10^9 eV.

G75.   D $N_A Z/A$ = (atoms/mole) × (electrons/atom) / (grams/mole) = electrons/gram. This is almost consistent for all materials except hydrogen.

2.4

G76.   D   Binding energy increases as Z increases, and with decreasing distance from the nucleus.

G77.   D   This is the definition of binding energy.

G78.   C   Binding energies are proportional to Z^2, thus: $E/69.5 = 8^2/74^2$

G79.   B   The binding energy increases approximately as the square of the atomic number.

G80.   B

G81.   C   The difference in mass between a deuteron and its constituents (1 proton + 1 neutron) is 0.00238 amu, or 2.22 MeV. This is the binding energy of the deuteron.

G82.   E   If the energy of the photon is less than the binding energy of the electron, ionization is impossible.

G83.　B　The binding energy per nucleon usually increases after radioactive decay, as the daughter nucleus is more stable than the parent. High binding energy implies stability.

G84.　A　Coulomb forces between protons are repulsive. The strong nuclear force, however, is always attractive, and occurs between neutron-neutron, proton-proton, and neutron-proton pairs. Large nuclei require additional neutrons to provide the extra binding forces necessary to overcome the coulomb forces of the protons.

G85.　B　Binding energy, or mass defect of a nucleus, is a measure of stability.

2.5

G86.　C　The maximum number in the outer shell is 8 (hence 8 groups in the periodic table) but in general the maximum no. of electrons in a shell of quantum number n is $2n^2$.

G87.　B　There can never be more than 8 electrons in the outer shell of an atom. Thus, there are 8 groups in the periodic table; elements with the same number of electrons in the outer shell are found in the same group. Shells other than the outer shell can contain twice the square of the shell's principal quantum no. (n) up to n = 4 (i.e., 2 for the K-shell, 8 for the L-shell, 18 for the M-shell, 32 for the N-shell).

G88.　C　All four elements have complete K- and L-shells of 10 electrons total. The M-shell contains a maximum of 18 electrons ($2n^2$), and would thus contain 6, 7, 8, and 9 electrons respectively for the four elements listed. The outer shell, however, contains a *maximum* of 8 electrons, meaning that potassium and argon both have 8 electrons in the M-shell, but the nineteenth electron in potassium goes into the N-shell. Thus argon is inert, while potassium is highly interactive.

3.1

G89.　A　Radioactive decay is represented by a straight line on a semi-log graph. B would be the shape of exponential decay on a non-logarithmic plot.

G90.　A　Activity (A) is defined as the number of nuclear transformations (dN) occurring in a given quantity of material per unit time (dt). A = dN/dt. Since dN/dt = λ N, A is also equal to λN.

G91.　D　The energy available for kinetic energy of the emitted particle is found by subtracting the masses of the emitted particle and final nucleus from the mass of the initial nucleus; the result is converted into energy units using E = mc^2. The mass of a nucleus is less than the sum of the masses of the constituent protons and neutrons by an amount called the "mass defect," which is greatest for the most stable nuclei.

G92.   C   After 0.5 half-lives, more than 25% of the activity has decayed. This can be seen by examining the shape of an exponential decay curve. The activity remaining is 70.7%.

G93.   D   The average life is 1.44 × half-life. The half-life is inversely proportional to the decay constant, and is unaffected by T or P.

G94.   E   A = A_0(initial) × exp (-0.693 × t/half-life) = 10 × exp (-0.693 × 72/12) = 0.156 mCi.

G95.   D   After n half-lives, the activity of a source is reduced to (1/2) to the power n, times its initial activity.

G96.   D   The decay constant, λ, is the fraction of radioactive atoms which will decay per unit time. In this case, λ = 0.01 per hr. The relationship between λ and half-life is: $T_{1/2}$ = 0.693 /λ. In this case: $T_{1/2}$ = 0.693/0.01 = 69.3 hrs.

G97.   D

G98.   C   In a time equal to half of a half-life, decay will be slightly more than one fourth, so 14 would be a good guess. Using the decay equation gives 14.14.
$A = A_o e^{-\lambda t}$, where $\lambda = 0.693/T_{1/2}$ = 0.693/6 hr, and t = 3 hr.

G99.   C   Combining physical and biological half-lives always gives an effective half-life shorter than either of the two components. $T_{eff} = (T_{bio} \times T_{phys}) / (T_{bio} + T_{phys})$

G100.   C   Although the initial dose rates may be similar, the total dose will be less. Total dose equals dose rate times average life. The average life equals 1.44 times the half-life.

G101.   B   The half-life is equal to 0.693/(decay constant).

G102.   C $$\frac{1}{T_{eff}} = \frac{1}{T_{phy}} + \frac{1}{T_{bio}}$$

3.2

G103.   E   Heavy nuclei tend to decay by alpha particle emission. Z decreases by 2, and A decreases by 4. An example is the decay of radium to radon.

G104.   B   When alpha decay occurs, A decreases by 4 (226 to 222). The gamma emissions for which radium sources are used clinically are due to the decay of Ra daughters further along the decay chain.

G105.   A   The mass number decrease of 4 indicates the ejection of an alpha particle only.

G106.   E   Alpha, example: $^{226}_{88}$Ra to $^{222}_{86}$Rn

G107.   B   Beta minus, example: $^{3}_{1}$H to $^{3}_{2}$He

G108.   D   Beta plus, example: $^{18}_{9}$Fl to $^{18}_{8}$O

G109.   C   Isomeric, example: $^{99m}_{43}$Tc to $^{99}_{43}$Tc

G110.   A   The mass number (A) does not change during isobaric or isomeric decay.

G111.   E   The total energy emitted is the difference between the ground states of the parent and daughter nuclides. There are three possible pathways.

G112.   C   A beta minus decay is always accompanied by an antineutrino, which shares the total beta-ray energy of the emission.

G113.   B   The decay scheme is an example of beta minus decay. The mass number (A) is unchanged.

G114.   B   The atomic number (Z) increases by 1.

G115.   A   Beta minus decay is always accompanied by antineutrinos.

G116.   A   Beta minus decay produces a continuous spectrum of electrons from 0 to E_{max}. All three spectra will include 0.511 MeV electrons.

G117.   B   The maximum beta-ray energy is 4.86 MeV.

G118.   A   X will decay in one of three ways. Each pathway will emit 4.86 MeV in a combination of beta-rays, gamma-rays and neutrinos.

G119.   B   Beta plus is possible because the atomic mass difference (M → M − 2 MeV) is greater than 1.022 MeV. Isobars have the same A. An isomeric transition implies that Z does not change.

G120.   D   Z changes, so an isomeric transition is ruled out. The transition is beta minus decay, in which Z increases by 1. Beta particles are emitted with a spectrum of energies; the available energy is divided between the beta particle and an antineutrino.

G121.   C   In beta decay, electrons are always emitted with a spectrum of energies. As in the case of cobalt-60, this decay may be accompanied by the emission of one or more monoenergetic photons.

G122.   B   A neutron changes into a proton plus an electron, which is ejected from the nucleus.

G123.   D   In beta decay, the kinetic energy released is shared between an electron, positron, and daughter nucleus. Although the sum of the energies of the three particles is always the same, each particle may have a spectrum of energies.

G124.   A   Molybdenum decays to the metastable state of technetium by beta minus decay. The decay of technetium from the metastable to the ground state releases a 140 keV gamma-ray, which is used in nuclear medicine imaging.

G125.   B   Gamma emissions can be 2.16, or 3.75 - 2.16 = 1.59 MeV. Beta minus spectra have maximum energies of 1.11, 1.11 + 1.59 = 2.70, or 1.11 + 3.75 = 4.86 MeV. Anti-neutrinos will also be emitted with maximum energies equal to the betas. Characteristic x-rays may also be emitted, but we cannot tell from the diagram what their energies will be.

G126.   D

G127.   B

G128.   B

G129.   C

G130.   A   In positron emission, the energy available for the positron and neutrino is the difference in energy levels – 2× the rest mass of an electron, i.e., 2.21 – 1.02 = 1.19 MeV.

G131.   D   The rest masses of the two particles combine to yield two gamma-rays of 0.511 MeV each.

G132.   E   Both electron capture and beta plus decay (which can compete in the same nuclide) reduce the number of protons by 1, and increase the number of neutrons by 1.

G133.   C   Positron decay is always accompanied by neutrino emission, and the positrons have a spectrum of energies. When the positron stops, it annihilates with another electron, producing a pair of 511 keV photons.

G134.   A   The mass equivalent of ^{99m}Tc is greater than that of ^{99}Tc by the transition energy of 140 keV.

G135.   B   1. ^{99m}Tc emits 140 keV gamma-rays. A scattered photon could have an energy of 130 keV.
            2. Characteristic x-rays resulting from a photoelectric interaction with the lead container could be detected.
            3. Cerenkov radiation is emitted when charged particles travel faster than the speed of light in a medium such as water.
            4. Auger electrons are emitted when characteristic x-rays are reabsorbed by the same atom. There is no such thing as an Auger x-ray.
            5. Annihilation radiation occurs when a positron and an electron combine, producing two 0.511 MeV photons.

G136.   C   Photon, (0, 0 MeV): isometric transition.

G137.   E   In beta decay and electron capture the daughter can be created in an excited state; decay to the ground state is then accompanied by the emission of the excitation energy in the form of one or more gammas. If the daughter has a measurable lifetime (e.g., ^{99m}Tc) the gamma emission is said to be an isomeric transition.

G138.   E   Energy is emitted as gamma-rays, or as conversion electrons resulting in characteristic x-rays and Auger electrons.

G139.   C   Characteristic x-ray (photon) (0, 0 MeV).

G140.   D   Electron capture, example: $^{51}_{24}$Cr to $^{51}_{23}$V

G141.   A   The nuclide decays in an isobaric transition (A → A) by either electron capture (Z → Z – 1) or beta plus (Z → Z – 1).

G142.   D   The transformation is an electron capture decay since Z → Z – 1 and a positron (e$^+$) is not emitted.

G143.   B

G144.   C

G145.   D

G146.   E   Positron emission and electron capture often occur in competition in the same isotope, when the number of neutrons is too low for stability. In electron capture, an electron, usually from the K-shell, combines with a proton to create a neutron and emitted neutrino. In filling the resulting K-shell vacancy, characteristic x-rays and Auger electrons are emitted.

G147.   A   Energy is transferred directly to an inner-shell electron, which is then ejected.

G148.   E   Internal conversion is an isomeric transition; energy may be emitted as a gamma ray, or transferred directly to an inner shell electron which is ejected. The vacancy is filled by an outer electron, and the energy difference emitted as characteristic x-rays or Auger electrons.

G149.   C   In the Auger process, one may think of a characteristic x-ray as being emitted, but immediately undergoing a photoelectric interaction with another orbital electron, which is then emitted from the atom. The energy of the Auger electron is equal to the energy of the characteristic x-ray less the binding energy of the electron. The Auger process is similar to internal conversion, except that in the latter case the fictitious photon precipitating the interaction is generated from the nucleus.

3.3

G150.   E   Cyclotrons can accelerate only charged particles; radioisotopes can be created by bombarding samples placed in the neutron flux of a reactor (e.g., ^{60}Co). Examples of radionuclides prepared by methods A-D are:
A: ^{137}Cs, ^{90}Sr
B, C: various positron emitters with short half-lives e.g., ^{11}C, ^{13}N, ^{14}O and ^{15}O.
D: ^{99m}Tc

G151.   A   Cyclotron-produced radioisotopes are typically short-lived positron emitters.

G152.   B   Radium is part of the uranium decay series.

G153.   E ^{99}Mo decays to ^{99m}Tc in a Tc generator. Transient equilibrium is established before elution of the Tc.

G154.   D ^{90}Sr is a beta emitter, used to make eye applicators and "check sources" to monitor ion chamber constancy.

G155.   C   Cobalt-59 is bombarded with neutrons to create cobalt-60.

G156.   B   Adding neutrons to the nucleus may result in too many neutrons for stability, leading to beta minus decay.

G157.   C   Too many protons in the nucleus result in decay by beta plus or electron capture.

G158.   C   A proton is added to the nucleus, and usually one or more neutrons are knocked out, resulting in too many protons for stability.

G159.   B   In general, the higher the atomic number, the larger the neutron-proton ratio for stability. Thus when a uranium nucleus splits into two intermediate-Z nuclei, these have too many neutrons, and thus will decay by beta minus.

G160.   C.   Positron emitters are required. These are usually short lived and proton rich. They are not easily made in a nuclear reactor, and an on-site cyclotron is usually used for production.

3.4

161.   B   Secular equilibrium occurs when the daughters' half-life is very much shorter than the parents' half-life. The decay constant (λ) is inversely proportional to the half-life ($\lambda = 0.693/T$). If λ_d is slightly greater than λ_p, T_d will be slightly shorter than T_p and secular equilibrium will not occur.

G162.   B   Transient equilibrium occurs when the daughters' half-life is somewhat shorter than the parents' half-life. It is only incidental that ^{99}Mo decays to a metastable state of ^{99}Tc.

G163. B In transient equilibrium, the activity of the daughter is slightly higher than the activity of the parent. The activity of ^{99m}Tc compared to its parent ^{99}Mo is somewhat less because some of the decay is to non-metastable states of ^{99}Tc.

G164. A See above.

G165. A See above.

G166. C The activity of the sample increases exponentially with time (the curve is the inverse of the decay curve). After one half-life, half the maximum activity (A_{max}) is achieved; after two half-lives, $0.75\ A_{max}$ is achieved, and so on. After n half-lives the activity is $(1 - 0.5^n)\ A_{max}$. Thus four half-lives give 94% A_{max}.

G167. D Examples of equilibrium are: ^{226}Ra to ^{222}Rn in a sealed container (secular), and ^{99}Mo to ^{99m}Tc (transient). The daughter decays with the longer half-life of the parent, since the rate of production of the daughter is equal to the rate of decay of the parent.

G168. B Secular equilibrium occurs when the parent has a long half-life. After about four half-lives of the daughter, the apparent decay of the daughter and parent are equal.

G169. C ^{99}Mo to ^{99m}Tc is an example of transient equilibrium. Although the half-life of the parent is longer than that of the daughter, the decay of the parent is significant during the time taken to achieve equilibrium (about four half-lives or 24 hrs).

G170. E Statements A-D are all false. When equilibrium is achieved, whether transient or secular, the activity of the daughter is slightly greater than the activity of the parent (except when there is more than one decay mode whereby less than 100% of the parent decays yield the daughter).

G171. E A and B are both true.

G172. C The isotope in question (radon) must be a decay product of another isotope with a much longer half-life, otherwise it would not be found at all. Thus, the isotope appears to decay very slowly, with the half-life of the parent.

3.5

G173. B Specific activity (Ci/g) = 40 Ci/10 g = 4 Ci/g. 1 Ci = 3.7×10^{10} Bq, so the answer is 1.48×10^{14} Bq/kg.

G174. B 1 Bq is equal to one disintegration per second (dps). 1 µCi is equal to 3.7×10^4 dps. $(7.4 \times 10^6$ Bq$)/(3.7 \times 10^4$ µCi$) = 2 \times 10^2 = 200$ µCi.

G175. B 1 Ci = 3.7×10^{10} Bq. Therefore, 10 mCi = 3.7×10^8 Bq = 3.7×10^2 MBq.

3.6

G176. C Exposure $(X) = [\Gamma \times A \times t]/d^2$
$X = ((2 \times 5 \times 3)/(100^2)) \times 1000(mR/R) = 3$ mR

G177. E Exposure rate = (Exp. rate const.) $\times$ activity $\times (l/d)^2$ where d and the exposure rate constant are both expressed in cm. Exp. rate $= 3.3 \times 10 \times (1/100)^2 = R/hr = 3.3$ mR/hr.

G178. C The exposure rate at r cm from A mg of Ra is: $8.25 \times A \times (1/r)^2$ R/hr
If A = 10 and r = 100 cm, this becomes 8.25 mR/hr.

G179. C Cobalt emits two gammas per disintegration (99.8% of the time). Half-life has no influence on gamma factor, and although cesium has a *lower* energy gamma than cobalt, this does not have a very marked effect on the gamma factor.

4

G180. B

G181. A The anode is the target (positive) and the filament is the cathode (negative).

G182. C Four rectifiers are required for full-wave rectification.

G183. E

G184. B Transformers are used in x-ray circuits to step-up voltage from volts to kilovolts required to generate x-rays, and also to step-down voltage (and therefore step-up current) in the filament circuit.

G185. E The milliammeter measures the current in mA. It is connected in series with the x-ray tube itself.

G186. A Rectifiers allow current to pass in only one direction; an x-ray tube acts as a rectifier in a "self-rectified" circuit.

G187. C In a transformer the ratio of primary to secondary voltage is in the ratio of the respective numbers of turns on the coils. V1/V2 = N1/N2.

G188. B Transformer oil is a good electrical insulator, and permits higher voltages to be achieved without electrical discharge.

G189. E The output of a step-up transformer is higher voltage, but lower current and lower power than the input.

G190. D Thermionic emission is the emission of electrons from the heated filament. A dual focus enables the focal size to be kept small except when a high power technique is used, when a larger spot allows the heat to be spread over a larger target area. A small target angle increases the ratio of actual to effective focal spot size. The rotating anode greatly increases the actual focal area, while maintaining a small effective focal spot.

G191. A The rotating anode spreads the heat load over a washer shaped disk rather than a single spot, thus increasing the heat load of the tube while maintaining a small effective focal spot.

G192. D One filament is used for low current exposures when a small focal spot will give a sharper image. At high currents, heat dissipation in the target is a problem if the focal spot is too small, so a second filament with a larger focal spot is used.

G193. C The process whereby a filament is heated to a sufficient temperature to emit electrons is called "thermionic emission."

G194. C Thermionic emission is the physical process by which electrons are emitted from the filaments of x-ray tubes, linear accelerators, klystrons, and magnetrons. The heat generated in (A) and (B) can be considerable but is counterproductive, while the heat generated in (D) is negligibly small.

G195. E The effective energy of an x-ray beam is approximately 1/3 to 1/2 of the kVp. This may be increased by additional filtration. X-ray energy produced by bremsstrahlung, is independent of the atomic number (Z) of the target material and of the mAs. Subject contrast is a function of x-ray energy.

G196. C Four rectifiers are required for full-wave rectification; the voltage ripple is 100% for both full- and half-wave rectification. The x-ray spectrum depends only on the kVp, filtration and the use of single- or three-phase power.

G197. D Three phase units have nearly constant voltage output resulting in higher beam current, average voltage, and dose rate.

5.1

G198. A A K_α x-ray is characteristic radiation emitted due to an electron transition from the L-shell to the K-shell. In order for a K_α x-ray to be emitted, a K-shell electron must be ejected from the atom. The incoming particle or photon must have at least as much energy as the binding energy of the K shell.

G199. A See answer to G198 above.

G200. B See answer to G198 above.

G201. B See answer to G198 above.

G202.   B   An Auger electron is an electron removed from an outer orbit of an atom by an x-ray (characteristic radiation) from an inner orbit, in this case, a transition from the L- to the K-shell. The kinetic energy of the Auger electron is equal to the difference between the x-ray energy and the binding energy of the electron. Therefore;
$$E_{x\text{-ray}} = 25.3 + 0.7 = 26 \text{ keV}.$$
The energy of the characteristic x-ray is equal to the difference of the binding energies of the K- and L-shell.
Therefore $E_{x\text{-ray}} = B.E._K - B.E._L$
$26 \text{ keV} = 30 \text{ keV} - B.E._L$
$B.E._L = 4 \text{ keV}$

G203.   B   Characteristic x-rays of tungsten are emitted from electrons accelerated to greater than 69 keV. Spectrum I is produced by 50 kVp electrons, spectrum II by 100 kVp electrons.

G204.   C   Photons below 10 keV have been removed by filtration by both spectra.

G205.   B   Spectrum II includes photons up to 100 keV.

G206.   D   The minimum photon energy from both spectra is about 10 keV.

G207.   D   Spectrum I does not have K characteristic x-rays. The K peaks appear on spectrum II.

G208.   A   The maximum photon energy from spectrum I is 50 keV which is produced by a potential of 50 kVp.

G209.   B   The exposure rate is a function of the intensity and the potential (kVp). The area under II is much greater than under I.

G210.   B   The HVL increases as the potential (kVp) increases.

G211.   E   Either spectra can be produced by single or three phase generators.

G212.   A   Scatter increases with kVp in the diagnostic range.

G213.   D   The energy of a characteristic x-ray is equal to the difference between the binding energies of the two shells involved in the transition.
$$E_K - E_L = (70 - 8) \text{ keV} = 62 \text{ keV}$$

G214.   A   For a tube voltage below 69.5 kV, there cannot be any K characteristic radiation, and L x-rays are normally filtered out. Above 69.5 kV the production of K x-rays increases with increasing kV. For a tungsten target, bremsstrahlung x-rays will account for the greater part of the spectrum.

G215.   C   The characteristic K x-rays at about 17.5 and 19 keV are desirable for maximum calcium and soft tissue contrast.

G216.   E   Most of the energy (99%) is lost as heat. Of the energy which is converted into x-rays, approximately 90% is bremsstrahlung and 10% is characteristic radiation, although this varies with the Z of the target material.

G217.   B   C-D are irrelevant. A is true only if the electron energy increases above the binding energy of one of the electron shells in the atom.

G218-222. BCBCB
              Possible characteristic x-rays are found by taking energy differences between shells (30, 29.3, 26, 4, 3.3, 0.7) given in keV. Bremsstrahlung can have any energy up to the maximum incident electron energy of 50 keV.

G223.   C   A, B, and E affect the shape of the spectrum, but not the maximum photon energy.

5.2

G224.   E   Most of the soft radiation in an x-ray beam is absorbed in the patient and does not contribute to the image. Hardening of the beam, by adding filtration, reduces the patient dose without appreciably reducing the radiation reaching the detector. The load on the tube, however, is increased. The field size is not affected. The quality (average energy) is increased resulting in greater scatter and wider overall latitude.

G225.   A   An x-ray beam output can be described by its quantity and quality. The quantity or intensity of the beam is best described by its exposure rate. The quality of the beam is best described by a graph of its intensity as a function of photon energy. The quality of the beam depends on the wave form, filtration and applied voltage (kVp).

G226.   A   The quality of the beam is independent of the tube current (mA). The quantity of the beam is proportional to mA.

G227.   B   Small changes in filament current will affect the quantity but not the quality of the beam.

G228.   A   See answer above. The graph of intensity versus photon energy is very difficult to measure. The HVL is usually used as a crude measure of the quality.

G229.   D   HVL = 0.693/linear attenuation coefficient.

G230.   D   A, B, and C determine the spectrum reaching the absorbers. For E, measurements made too close to the absorbers or with a large beam size may enable scatter to reach the detector, giving incorrect values.

G231.   B   For a monochromatic beam, two half-value layers reduce the intensity to 25%, but for a broad spectrum the added absorbers harden the beam, increasing the penetrability; the second HVL is greater than the first.

G232.   D   Two HVLs reduce the intensity to 25%, but the reduction due to the inverse square also applies: 25% times $(50)^2/(70)^2 = 13\%$.

G233.   E   The value of the HVL depends on both the kVp and the amount of filtration in the beam, and is more easily measured than either of those. Filtration requirements are specified in terms of minimum HVLs for various kVps.

G234.   D   Full-wave rectification doubles the effective tube current (as compared to half-wave rectification) without changing the effective photon energy. Decreasing filtration increases the dose rate, but decreases the effective photon energy.

G235.   E   Removing filtration from a beam increases the number of low energy photons thus decreasing the HVL. Since the number of photons in the beam increases, the dose rate increases.

6.1

G236.   A   Beta-rays (positive or negative) are ionizing particles.

G237.   D   Heat radiation consists of non-ionizing photons with energies of about 10^{-2} to 10^{-3} eV. ← *important*

G238.   D   Visible light consists of non-ionizing photons with energies of about 1.8 to 3.0 eV.

G239.   C   X-rays and gamma-rays are ionizing photons of energies above about 10 keV.

G240.   E   Ultrasound is a mechanical motion that is propagated as a wave through a medium.

G241.   D   Sound waves of any frequency are not ionizing. Electromagnetic and particulate radiation with sufficient energy to remove an electron from an atom are classified as ionizing radiation. Neutrons are indirectly ionizing, as they release "knock on" protons.

G242.   D   X-rays have photon energies in the high keVs, light in the low keVs, radiowaves 10^{-3} eV or less; ultrasound does not consist of photons.

in order of wavelengths ↓
radiowaves
ultrasound
light
x-rays

G243.   A   Radiowaves have wavelengths in meters or cm, light in 10^{-7} cm ultrasound in mm; x-rays have shorter wave lengths than light.

G244.   D   X-rays, since frequency is proportional to energy.

G245.   C   The speed of sound varies with the medium, but is about 1500 meters per second in tissue, while radio waves, x-rays and light travel at 3×10^8 meters per second.

G246.   C   The electromagnetic spectrum consists of (from longest to shortest wavelengths): radio, micro, infrared, visible, ultraviolet, x- and gamma-rays. Energy increases as wave length decreases.

6.2

G247.   D   Electromagnetic radiation, unlike charged particles, is not deflected by magnetic fields.

G248.   D.   X-rays originate from electron shells of an atom, while gamma-rays originate from the nucleus.

G249.   B   Energy is inversely proportional to wavelength. Thus
$E_{visible} = 1.2 \times 10^{-10}/6 \times 10^{-5}$ MeV = 2 eV.

G250.   E   Using the inverse square law: $I[75] = I[50] \times (50/75)^2 = 4.4$ mR/min.

G251.   E   By the inverse square law, 100 mR$/I_2 = (100)^2/(50)^2$.

G252.   C   frequency = velocity / wavelength.
For electromagnetic radiation the velocity (c) = 3×10^8 m/s
frequency = $3 \times 10^8 / 6 \times 10^{-2} = 5 \times 10^9$ Hz.

G253.   C   velocity of light (const.) = wavelength × frequency
energy = Planck's constant × frequency

7.1

G254.   A   In the interaction of radiation and matter, energy is transferred to electrons and photons. The energy transferred to photons is called scatter. The energy transferred to electrons is dissipated locally in collisions with atoms and is called absorption. Absorption takes place in all attenuation processes.

G255.   B   See above.

G256.   D   Absorption normally decreases with increasing photon energy, and increases with increasing atomic number.

G257.   A   From approximately 1-10 MeV photon energy, only Compton effect is important, and the total cross section is a minimum. The exact energy at which the minimum total cross section occurs depends on the atomic number of the absorbing material.

G258.   B   The fraction of the photons transmitted by x cm of an attenuator is I/I_o where I is equal to the number of transmitted photons through x cm of the attenuator, I_o is the number of photons in the unattenuated beam and μ is the linear attenuation coefficient. The fraction of the photons attenuated in x cm is equal to $1 - I/I_o$.

G259. D The fraction of the beam transmitted is I/I_o where:
$I/I_o = e^{-\mu x}$, $\mu = 0.0693$ cm^{-1}; $x = 10$ cm
$I/I_o = e^{-0.693} = 0.50$
Another approach is to calculate the HVL (50% transmission):
HVL $= 0.693/\mu = 0.693/(0.0693$ cm$^{-1}) = 10$ cm.

G260. D The linear attenuation coefficient depends on beam energy and attenuating medium. The formula is used to calculate the intensity of a beam after it has passed through x cm of an attenuating medium.

G261. C The mass attenuation coefficient is the linear attenuation coefficient divided by the density of the material. Whereas the probability of an interaction is proportional to Z cubed for the photoelectric effect, and Z for pair production, it is independent of Z for the Compton effect, and all materials have an equal probability, gram for gram (since all materials except H have almost equal numbers of electrons per gram). The mass attenuation coefficient represents the attenuation per unit mass of a material.

G262. D The intensity of the transmitted x-ray beam is given by:
$I_x = I_o \exp(-0.693\ x/\text{HVL})$ and $I_x = I_o \exp(-\mu x)$, therefore, HVL $= 0.693/\mu$
and if $x =$ HVL: $I_x = I_o \exp(-0.693) = I_o/2$.

G263. E HVL $= 0.693/$linear attenuation coefficient $= 0.693/0.05 = 13.86$.

G264. A The mass energy absorption coefficient is equal to the linear absorption coefficient divided by density. The mean attenuation length is equal to the linear absorption coefficient times 1.44. There is no such thing as a scatter coefficient.

G265. E $e^{-0.693n}$ is equal to the fraction of photons transmitted. Therefore $(1 - e^{-0.693n})$ is equal to the fraction absorbed.

G266. C CT$_{number} = 1000 \times [(\mu_{material} - \mu_{water})/\mu_{water}]$, where μ is the linear attenuation coefficient.

G267. D Photoelectric absorption is proportional to Z^3/E^3.

7.2

G268. D The second half-value layer is the amount of an attenuating material that will reduce the intensity of a photon beam from 50% to 25%. The first HVL is approximately 2.4 mm Al. The second HVL is approximately 3.6 mm Al.

G269. C The exposure must be reduced from 142 mR (1 mm Al added) to 71 mR. The total thickness of Al required necessary is a little less than 4 mm. The HVL with 1 mm Al added is therefore a little less than 3 mm (4 – 1). The closest answer is 2.9 mm Al.

G270. B Technetium-99m is a monoenergetic gamma-ray emitter. The first and second HVLs would be approximately the same.

G271. B Three TVLs will reduce the exposure by a factor of 1000 (10^3). Ten HVLs will reduce the exposure by a factor of 1024 (2^{10}).

G272. B The primary radiation is reduced by 2 TVLs. Half of the attenuating material will reduce the dose by one TVL (10%). The total dose, however, will include the contribution of scattered radiation which will depend on the energy of the radiation.

G273. C The activity is increased by a factor of 4 (1 to 4 Ci). Therefore two additional HVLs are required.

G274. A Passing through the first HVL hardens the beam, so a greater thickness of material is required to reduce the intensity by a further 50%.

G275. E 6 cm = 5 HVLs. The transmitted fraction = $1/2^5 = 1/32 = 0.03125 = 3.1\%$.

G276. E As the beam passes through the Al in which the HVL is measured, it is filtered and hardened. Therefore, more Al is required to reduce the beam's intensity from 50% to 25% than from 100% to 50%.

G277. A Adding filtration to a beam selectively filters out the lower energy photons, leaving a more penetrating beam with a greater HVL (but a lower intensity).

G278. B With broad beams, the detector records scattered as well as primary radiation, which incorrectly increases the apparent HVL.

G279. A For a polyenergetic beam, the first HVL hardens the beam, making the second HVL greater than the first.

7.3

G280. D The higher atomic number of calcium results in greater absorption by the photo-electric process.

G281. A The absorption of x-rays (dose) can be calculated by multiplying the exposure by the f-factor which is a function of the ratio of the mass attenuation coefficients of the material and air. The mass attenuation coefficient is a function of the atomic number (Z) of the material and the energy of the incident photon. Iodine has a higher Z than bone and will absorb more than bone.

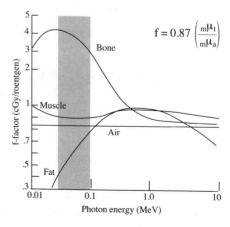

G282.   A   See answer to G281 above.

G283.   B   See answer to G281 above.

G284.   A   See answer to G281 above.

G285.   B   See answer to G281 above.

G286.   B   Photoelectric: considered total absorption, since the characteristic x-rays are self-absorbed in the attenuating material, in a very short distance.

G287.   B   A photon disappears following a photoelectric interaction. Following a Compton interaction, the scattered photon may interact again. The photon also disappears following pair production.

G288.   B   See answer above.

G289.   B   See answer above.

G290.   A   The incident photon is totally absorbed in the photoelectric interaction.

G291.   B   The probability of a photoelectric interaction decreases rapidly ($1/E^3$) with photon energy.

G292.   D   The photoelectric component (t) of the linear attenuation coefficient is proportional to Z^3 and the physical density (ρ). $(t_B/t_A) = (14/7)^3 \times (2/1) = 2^3 \times 2 = 2^4 = 16$.

G293.   B   This is an example of photoelectric absorption, since the product is an 80 keV photoelectron.

G294.   A   Characteristic radiation will be emitted as a result of an electron from an outer shell falling into the K-shell vacancy.

G295.   A   Auger electrons are electrons produced by characteristic x-rays interacting with electrons in the same atom from which the characteristic x-rays are emitted. The probability of producing Auger electrons at relatively low x-ray energy (10 keV) is high.

G296.   A   Electrons interact because of their charge and mass by ionization and excitation along their paths. In general, the path of an electron in any medium, except air, is very short. (A 2 MeV electron has a range of 1 cm.)

G297.   C τ/ρ is proportional to Z^3. $0.004 \times (82/8)^3$ is approximately 4.

G298.   A   In the photoelectric interaction, if the incident photon energy is greater than the binding energy of the electron, characteristic x-rays may be produced.

G299.   B   Interactions can occur only with tightly bound electrons.

G300.   B   The energy of the incident photon must be greater than the binding energy of the electron, and the probability is greatest when the photon energy is a little greater than the electron binding energy.

G301.   A

G302.   A   In tissue, most of the energy is transferred to the kinetic energy of the ionized electron. The electron interacts with the surrounding atoms by ionization and excitation and loses all of its energy within a short distance from the point where it was emitted.

G303.   D   In a photoelectric interaction the photon is completely absorbed by the electron, which is then emitted from the atom with kinetic energy equal to the difference between the incident photon energy and the binding energy of the electron. The vacancy in the shell is then filled with an electron, causing a characteristic x-ray to be emitted.

G304.   C   Compton is the most probable interaction between 25 keV and 25 MeV.

G305.   B   Photoelectric interactions are more likely at low than high energy; after passing through a filter, the total beam intensity is reduced, but the beam contains a relatively greater number of high energy photons than before filtration.

G306.   E   Photoelectric absorption is proportional to Z^3/E^3. Choice (E), with $Z = 53$ and $E = 30$ has the highest Z/E ratio, and hence the highest photoelectric absorption.

7.4

G307.   A   No energy is lost in classical scattering.

G308.   E   Coherent interaction: scatter but no absorption.

G309.   C   Compton: scatter and absorption.

G310.   A   After a Compton interaction the scattered photon may interact again.

G311.   A   See answer above.

G312.   A   See answer above.

G313.   A   The wavelength of the scattered photon is always greater than the wavelength of the incident or primary photon in the Compton process, since its energy is reduced.

G314.   A   The primary photon is scattered without loss of energy in a coherent interaction.

G315.   B   The probability of Compton scatter (per gram) is about the same for Sn (Z = 50) and Al (Z = 13). However, since the probability of the photoelectric interaction is proportional to Z^3 per gram, it is more likely that the interaction with Sn will be by the photoelectric effect.

G316.   B   The energy of the incident photon is divided between the scattered photon and the recoil electron. The scattered photon may travel in any direction. In a coherent interaction the scattered photon carries off all the energy.

G317.   B   The electron may acquire any kinetic energy (KE) from 0 up to a maximum (E_{max}) which may be calculated by:
$E_{max} = E_{h\upsilon} - E_{h\upsilon\ min} = h\upsilon\ [1 - 1/(1 + 4E_{h\upsilon})]$
where $h\upsilon$ is the energy of the incident photon and $h\upsilon_{min}$ is the energy of the scattered photon.

G318.   A   The energy of the incident photon is divided between the scattered photon and the recoil electron. The minimum scattered photon energy is when the photon is scattered at 180° (backwards). In this case, the electron acquires its maximum kinetic energy and travels in the forward direction.

G319.   B   A neutrino is not involved in the interaction of radiation and matter; it is emitted during beta decay.

G320.   D   In a Compton interaction the photon loses some of its energy to the electron, which is then emitted at an angle of between 0° and 90° to the direction of the incident photon. The photon with reduced energy can be emitted at any angle, although the angles involved are related to the initial and final photon energies.

G321.   C   Even at low kV, coherent scatter contributes only a small part of the total scatter. The characteristic x-rays created by photoelectric interactions within the patient are of very low energy (because of the low Z of tissue) and have an extremely small range. Compton and photoelectrons also have a short range and are unlikely to leave the patient's body.

G322.   D   In a Compton interaction, the angle at which the scattered photon has the lowest energy is always 180°, i.e., backscatter. However, at 100 keV, the backscattered photon retains about 70% of the incident energy, while at 1 MeV the backscattered photon retains only about 20%. A rule of thumb for megavoltage photons is that backscatter is about 250 keV, and side scatter is about 500 keV.

G323.   C   The probability of a Compton interaction decreases only slightly with increasing photon energy.

G324.   A   Coherent scatter, also known as Raleigh or unmodified scatter, occurs only for very low energy x-rays, and is of little concern in radiology.

G325.   C   The probability of a Compton interaction depends on the number of electrons per gram, which is roughly the same for all elements.

G326.   C

G327.   D   Pair production increases slowly with increasing energy, and may become the
              dominant process in high-Z materials at multi-MeV energies.

G328.   C   The number of Compton interactions depends on the number of electrons present.
              Most materials have the same number of electrons per gram, but hydrogen is an
              exception. It has one electron per nucleon (proton), whereas all other atoms have
              approximately one electron to every two nucleons (proton + neutron). Thus
              hydrogen has approximately twice as many electrons per gram as does water. Only
              the absolute (not relative) number of interactions depends on photon energy.

G329.   A   Secondary photons can be backscattered, but secondary electrons cannot.

G330.   B   At high photon energies backscattered Compton photons always have an energy of
              255 keV. When the photon is backscattered the electron must go in the forward
              direction: 0°.

G331.   E

7.5

G332.   A   Positron, $(+1, 0.511$ MeV$)$: it is very unstable, and will annihilate by combining with
              an electron and creating two 0.511 MeV photons.

G333.   B   The threshold energy for pair production is the energy necessary to create an
              electron pair, an electron and a positron. $2 \times m_e c^2 = 1.02$ MeV. The rest mass
              $(m_e c^2)$ of an electron or positron is 0.511 MeV.

G334.   A   In pair production an incident photon disappears and an electron pair (electron and
              positron) is produced. A negative (electron) and a positive (positron) particle are required
              in order to conserve charge.

G335.   B   The total energy of an incident photon is divided among the kinetic energies of the
              pair and the energy necessary to create the pair (1.02 MeV). A negligible amount of
              energy is transferred to the atomic nucleus.

G336.   B   A positron cannot exist in nature. It interacts with an electron and both particles are
              annihilated. The interaction produces two 0.511 MeV photons that must travel in
              opposite directions in order to conserve momentum.

G337.   B   The electron and positron are not restricted in direction; however, the annihilation
              photons created when the positron-electron pair annihilate are emitted in opposite
              directions.

G338.   B   1.02 MeV is emitted as two 0.51 MeV photons.

G339.   A   The energy of the initial photon, less 1.02 MeV (the rest mass of the two electrons created) is shared between the electron and positron. When the positron stops, it annihilates with another electron, producing a pair of 511 keV photons.

7.6

G340-344. EDCAB
Curve #5 (pair production) is zero below 1.02 MeV, and increases more rapidly in lead (because of lead's higher Z) than in water. Curve #4 (Compton) is almost equal in water and lead (when measured per gram, as opposed to per cm). The lead (curve #1) cross section is higher than that of water (curve 2) at low energies because the photoelectric effect is proportional to Z^3, and at high energies because pair production is proportional to Z. At intermediate energies where the Compton effect dominates, the two curves are almost equal. The photoelectric effect in water (curve 3) decreases very rapidly as energy increases.

G345.   A   The iodine K-edge is at 33.2 keV. X-rays with energy just above the K-edge are most likely to undergo a photoelectric interaction.

G346.   B   For muscle, a Compton interaction is most likely between about 25 keV and 25 MeV.

G347.   B   See above.

G348.   A   Photoelectric is more likely than Compton for fat at energies below 20 keV.

G349.   B   Photoelectric is more likely than Compton for bone at energies below 40 keV.

G350.   A   The K-edge of lead (Pb) is at 88 keV.

G351.   A

G352.   A

G353.   D

G354.   E

G355.   A   The mass attenuation coefficient is related to the probability of a photon interaction. When the photoelectric effect occurs, the probability is proportional to Z cubed; the probability of a Compton interaction is independent of Z. The net change in the mass attenuation coefficient, therefore, depends on the relative numbers of each type of interaction, which in turn depends on the photon energy.

G356.   B   Photoelectric absorption decreases rapidly as a function of photon energy, and predominates at low energies.

G357.   C   Compton scattering decreases only slowly with energy, and predominates at higher energies.

G358.   B   Photoelectric absorption depends on Z^3.

G359.   D   The values plotted are the mass absorption coefficients, i.e., per gram, rather than per cm. Had they been plotted as linear attenuation coefficients, a higher value in the Compton range would indicate a greater density.

8

G360.   B   Alpha particles have ranges in microns; for 8 MeV, about 95 µm. A 2 MeV beta particle has a maximum range of about 1 cm.

G361.   D   The electron's rest mass is low compared to its kinetic energy, so it must be traveling at nearly the speed of light. The alpha has the greatest total energy.

G362.   C   Only 2 is false. Photoelectric effect is a photon interaction, not an electron interaction.

9

G363.   D   H & D curve: optical density of film versus dose.

G364.   C   Optical density is the logarithm of the incident intensity divided by the transmitted intensity. Optical density is therefore additive. Thus, 1.5 + 1.5 = 3.0 is the optical density of the sandwich. The antilog of 3 is 1000, meaning that 1/1000 or 0.001 of the incident light is transmitted.

G365.   D   Optical density is the logarithm of the incident intensity divided by the transmitted intensity. Optical density can only be related to dose if the H & D curve is known.

G366.   B   Compton scattered photons travel in random directions and thus contain no useful diagnostic information. The grid absorbs most of these photons, thus improving image contrast.

G367.   D   Compton scattered photons travel in random directions and thus contain no useful diagnostic information. The grid absorbs most of these photons, thus improving image contrast. Photoelectrons rarely travel far enough to leave the patient.

G368.   C   The probability of a photoelectric interaction is proportional to the cube of Z (atomic number), whereas it is independent of Z for Compton interactions. Thus, a small difference in Z between different materials causes a large difference in the number of interactions which occur at low energies where photoelectric interactions are most common.

G369. C Collimating the beam decreases the area of exposure, using screens and a high kV technique both reduce patient dose. Grids improve contrast by "cleaning up" scatter, but require a somewhat higher dose to compensate for attenuation by the grid.

10

G370. A The standard deviation for the count rate $(\sigma_R) = (N)^{1/2}/t$. The average counts per sample are: 1000 cpm × 10 m = 10,000 counts.
$\sigma_R = (10,000)^{1/2}/10 = 100/10 = \pm 10$ counts.

G371. C The standard deviation (s) is equal to the square root of the total number of counts collected (N). The percent standard deviation (%s) is equal to (s/N) × 100.

$$2 = \frac{\sqrt{N}}{N} \times 100 = \frac{100}{\sqrt{N}}$$

Thus, $\sqrt{N} = 50$, and N = 2500 counts

G372. C If a large number of measurements are made approximately 67% will fall between $\pm\sigma$, and 96% will fall between $\pm 2\sigma$. $\sigma = \sqrt{2500} = 50; 2\sigma = 100$. Therefore, about 96% of a large number of repeated measurements will fall between 2400 and 2600.

G373. B The poisson distribution is an approximation of the binomial distribution in which the occurrence or nonoccurrence of the event is recorded. The normal distribution occurs when the variations are due to a large number of variables.

G374. A Both the poisson and the normal distribution occur because of random variations.

G375. A The photon distribution across an x-ray film depends only on whether the photons are absorbed or not absorbed.

G376. A Radioactive decay depends only on whether the atoms disintegrate or do not disintegrate.

G377. A $\sigma = (N)^{1/2}$ where N is the total number of events. Example: N = 10,000; $\sigma = 100$.

G378. B $\%\sigma = [(N)^{1/2}/N] \times 100$. Example: See example above, $\%\sigma = 1\%$.

G379. D The net count rate = $[(N_s/t_s) - (N_b/t_b)] = [(900/1) - (100/10)] = 890$ count/min.
The net $\sigma = [(N_s/t_s^2) + (N_b/t_b^2)] = [(900/1^2) + (100/10^2)]^{1/2} = 30$.

11

G380. A The mainframe is a physical (hardware) device that contains the CPU and main memory. All other hardware (terminal, input/output, digitizer, printer, plotter) are peripheral devices.

G381.　B　ROM is *read only memory* as opposed to RAM, *random access memory*. Programs may be put in or out of ROM and used but not revised. Programs may be put in and out of RAM and used and revised.

G382.　D　A byte is a combination of bits (usually 8 or 16) treated as a unit which stores one unit of information.

G383.　D　Only the RAM (random access memory) is erased when the computer is turned off. The other devices are either permanently (ROM) or semi-permanently magnetized.

G384.　D　The following table lists typical storage capacity in kilobyte (kB), megabyte, (MB) and/or gigabytes (GB) (1 byte = 8 bits)

　　　　　　O — Optical disk (1000 to 2000 MB) (or 1 to 2 GB)
　　　　　　M — Multiplatter hard disk 300 to 700 MB
　　　　　　T — Magnetic tape (1600 BPI, 2400 ft) = 46 MB
　　　　　　H — Hard disk 40 MB to 2.5 GB
　　　　　　F — Floppy disk 400 kB to 1.5 MB

G385.　D　A computer with a word size of 16 bits can address a maximum of 2^{16} locations. $2^{16} = 65,536$.

G386.　A　An input/output device is used for entering or extracting information from the computer while a storage device is used for storing data.

G387.　B　See answer to G386 above.

G388.　B　See answer to G386 above.

G389.　A　See answer to G386 above.

G390.　B　See answer to G386 above.

G391.　B　ROM stands for read only memory — any storage medium to which data cannot be written by the system in which it exists. Used for storing instructions that are used repetitively without modifications.

G392.　A

G393.　B　Real time pertains to the processing of events as they occur in clock-time.

G394.　A

G395.　B　A binary digit is a bit. A byte is a set of adjacent binary digits (bits) operated on as a unit by the computer. The most common size byte contains 8 bits.

G396.　A

G397.   A

G398.   A

G399.   B   A bit is one unit, or binary switch. A byte is composed of 8 bits, and a word is 2, 4, or more bytes, depending on the computer. A block is 512 bytes.

G400.   A   Booting a computer means reloading the operating system after the computer has been turned off. The contents of random access memory, however, may be lost when electrical power is turned off, and thus cannot be used to store the boot instructions.

G401.   D   A bit is one unit, or binary switch. A byte is composed of 8 bits. If all 8 bits are on, then the byte = 11111111 binary, or 255 decimal.

G402.   C

G403.   B   Parallel processors divide a task into pieces, with each piece being solved on a separate processor to increase computing speed.

12.1

G404.   E   The average annual natural background radiation in the United States is about 100 mrem (excluding radon). About 40 mrem/yr is contributed by radioactive material within the body, mostly ^{40}K. The rest is about equally divided between contributions from cosmic rays and radioactive materials in the earth. The other answers are examples of artificially produced radiation.

G405.   D   Millions of diagnostic x-ray procedures are done every year. The next most significant source of radiation dose to the population as a whole is from nuclear medicine exams.

G406.   C   According to BEIR III, the atomic bomb studies indicate that the lifetime risk is about 1 in 10,000.

G407.   C   Report No. 94 of the NCRP estimates an effective dose equivalent of 200 mrem (2 mSv) per year from radon, compared to 100 mrem (1 mSv) per year from all other natural sources. The dose equivalents from man-made sources are:

From NCRP Report #93:	mrem/year	mSv/year
Diagnostic radiology	39	0.39
Nuclear medicine	14	0.14
Nuclear power	0.05	0.0005

The effective dose equivalent from nuclear weapons testing has always been estimated at less than 1 mrem (0.01 mSv) and is decreasing yearly.

G408.   C   Radon decays through a chain of daughter products which may be inhaled, some of which decay by alpha emission.

G409.   E   NCRP #91 suggests 20, unless the energy spectrum of the particles is known.

G410.   C   Radon is a radioactive gas emitted primarily from rocky soils and/or building materials which contain trace amounts of uranium and radium ores. Radon concentrations in the air can reach hazardous levels in poorly ventilated basements and ground floors, particularly in certain parts of the country where soil and rocks contain large amounts of these ores.

12.2

G411.   E   See NCRP Report #116.

G412.   C   10 CFR Part 20 recommends that personnel monitoring shall be performed for occupationally exposed individuals for whom there is a reasonable possibility of receiving a dose exceeding one-tenth of the applicable MPD. It specifically excludes the use of monitors when the individual is "exposed as a patient for medical or dental reasons."

G413.   B   Selective filters are used in film badge holders in order to crudely distinguish between different types and energies of ionizing radiations. An open window (no filter) is used for beta-ray detection. Film badge reports should be considered with caution because of the difficulty of interpreting mixed-radiation exposures.

G414.   B

G415.   A

G416.   A

G417.   B

G418.   D   Film badges cannot measure exposures less than about 20 mR accurately. Various metal filters allow an estimation of the proportion of dose due to x-rays of different energies.

12.3

G419.   C   Exposure rate = (Exp. rate const) × activity × (l/d^2) = 12.9 × 10 × 1/10000 R/hr = 12.9 mR/hr. To reduce 12.9 to 2 requires 3 HVLs.

G420.   A   All walls must be considered primary barriers because there are no restrictions on the orientation of the beam.

G421.   B   In general walls are assigned use factors (U) of 1/4 or 1/16. The walls could not *all* have a U of 1 at the same time.

G422.   B   For areas such as corridors or toilets recommended occupancy factors (T) are 1/4 or 1/16. Occupancy factors are based on the time that an individual is exposed to radiation in the area. A busy, one person office could be assigned a T of 1.

G423.   B   Controlled areas are allowed levels of 0.1 rem per week (approximately 5 rem per year). Non-controlled areas are required to be shielded to levels of 0.002 rem per week (0.1 rem per year). In practice, most areas are shielded to well below the maximum permissible levels.

G424.   A   This is an additional restriction on non-controlled areas in order to insure that any combination of use and occupancy cannot raise the total dose to greater than the MPD.

G425.   D   The occupancy factor (T) is the fraction that the exposure should be decreased in order to correct for the degree of occupancy. Since T is increased by a factor of 8 and the activity (A) is doubled, the exposure is increased by a factor of 16. 4 HVLs ($2^4 = 16$) of lead are required to maintain the same radiation level. 0.3 mm × 4 = 1.2 mm Pb.

G426.   B   The weekly exposure at the console is proportional to $W \times U \times T / d^2$ where T = occupancy factor (always 1 in a controlled area); W = workload (doubled); U = use factor (the same). The exposure is therefore doubled, requiring one extra HVL.

G427.   C   This is hard to estimate exactly, so standard fractions are generally used for walls and floor.

G428.   B   If that space could be permanently occupied, i.e., a desk could be put there, then T is given the value 1. Otherwise, various fractional values are used.

G429.   C   Increasing the room size increases the distance to adjacent areas. Since dose rate decreases approximately as $1/R^2$ the required shielding decreases as the room size increases.

G430.   D   The maximum permissible dose recommended by NCRP #116 for nonradiation workers is 0.1 rem per year, if this is continuous exposure. Adding one tenth-value layer of shielding will reduce the dose from 1 rem to 0.1 rem.

G431.   B ^{32}P emits a spectrum of beta minus electrons with a maximum at 1.7 MeV which upon striking lead will produce high energy bremsstrahlung. A 0.5 mm Pb apron would not stop these x-rays.

G432.   A   Pb aprons are most effective at these x-ray energies. A 0.5 mm Pb apron will reduce the exposure by a TVL or more.

G433.   B   The 662 keV gamma-ray of ^{137}Cs would require 6.5 mm Pb to reduce the exposure by half.

G434.   B   The 740 keV gamma-rays of ^{99}Mo would not be effectively attenuated by the Pb apron. Even the ^{99m}Tc 140 keV gamma-ray with an HVL of 0.3 mm Pb would only be partially attenuated.

G435.   A   The approximately 30 keV photons of ^{125}I would be very effectively reduced by a 0.5 mm Pb apron.

G436.   B   The 511 keV annihilation photons are not effectively stopped by a 0.5 mm Pb apron.

12.4

G437.   B   A NaI well counter is an efficient measuring device for low-level gamma detection. It also provides discrimination for gamma-ray analysis.

G438.   A   A liquid scintillation counter allows 50% to 70% efficiency for measuring small quantities of the low energy beta-rays from tritium.

G439.   C   A GM counter has a fast response and the ability to detect low levels of gamma-rays.

G440.   E   An ionization chamber survey meter is capable of accurate x-radiation measurements with minimal energy dependence.

G441.   D   A TLD dosimeter is made of a material that stores energy when exposed to ionizing radiation. The energy is released as visible light when it is heated in a TLD reader. The light is detected by a photomultiplier tube. The small size and relative energy independence of the dosimeter makes it a useful personnel monitoring device.

G442.   C   Quenching gases (containing polyatomic molecules) are used in Geiger-Mueller counters to "quench" the avalanche of charge released by the passage of the initial charged particle. This decreases dead time, and prevents the total paralyzation of the counter.

G443.   E   Too high a voltage will cause conduction in gasses or recycling in a Geiger tube.

G444.   C   Proportional counters operate in this region.

G445.   B   In region A, some ions recombine before being collected. With sufficient voltage, all are collected. In region C, additional voltage causes multiplication.

G446.   D   Any event will trigger an avalanche, giving pulses of uniform size.

G447.   B

G448.   D   The high atomic number of iodine makes it highly sensitive to low energy radiation due to photoelectric absorption.

G449.   A   The cascade effect amplifies a weak signal.

G450.   C   Radiation produces ionization in the chamber gas which is measured as an electric current. The measured current (or integrated charge) can be used to determine the number of roentgens or the dose, but neither of these quantities can be measured directly.

G451.   C   The temperature correction factor or TPC is given by:
$$TPC = [(273 + t)/(273 + 22)] \times 760/P = 291/295 \times 760/750 = 1.0$$

G452.   D   Electron equilibrium is reached at approximately the d_{max} depth, or where the kerma and the dose are exactly equal.

G453.   C   1 roentgen $= 2.58 \times 10^{-4}$ C/kg of air, and can be measured absolutely with a free air chamber. All the other methods must be compared with a calibrated standard.

G454.   A   Each ionization event in a Geiger counter generates a single large pulse, making it very sensitive to low levels of radiation. Geiger counters cannot, however, measure dose directly.

G455.   D   Calorimetry is difficult to use in practice, so is not routinely used.

G456.   C   If the collection voltage is not high enough in an ionization chamber, recombination of ions can occur, resulting in a low reading.

12.5

G457.   D   Very low concentrations of radioactive materials when ingested can produce high localized radiation doses to internal organs.

G458.   D   3 years is 36 months, or approximately 18 half-lives.
The activity remaining is approximately $25 \times [1/2]^{18} = 9.5 \times 10^{-5}$ mCi
The exact value is: $25 \exp -(0.693 \times 3 \times 365 / 60) = 8 \times 10^{-5}$ mCi.
The exposure rate at 1 cm is: $1.45 \times 8 \times 10^{-5}$ R/hr $= 0.12$ mR/hr. The photon energy is about 35 keV. This exposure rate is so low that it is neither a hazard to staff, nor could it affect a radiograph.

G459.   D   When an unshielded container does not give any reading above background as measured with a sensitive Geiger survey meter, the contents may be disposed of as non-radioactive material.

12.6

G460.   D   In a controlled area, workers wear personnel monitors, such as film badges. If members of the public might occasionally be present, then the dose they might receive must not exceed 20 µSv (2 mrem) in any one hour. Note: this does not mean that the instantaneous dose rate cannot exceed 2 mrem/hr; the dose rate of the unit, the use factor and the workload must all be considered.

G461.   A   See 10 CFR Part 20.

THERAPY

Questions
&
Answers

THERAPY Contents

Brachytherapy

Radiation Protection 233

THERAPY Photon Dosimetry

HVLs in AL, then start defining in Cu

PD1. Definitions

T1.   The greatest backscatter factor in soft tissue is associated with:
   A. 30 kV$_p$ x-rays ✓ *(most PE, no scatter)*
   B. 4 MV linear accelerator x-rays *regardless of forward scatter*
   C. cobalt-60 teletherapy γ-rays
   D. 2 mm Al HVL x-rays *(mostly PE, no scatter)*
   (E.) 1 mm Cu HVL x-rays → *BSF peaks in soft tissue*

   BSF   maximum BS occurs @ approx .7mm Cu HVL

T2.   Tissue-maximum ratio (TMR) depends on:
   A. energy, SAD, depth and field size
   B. energy, SAD, and field size
   C. SAD, depth and field size
   (D) energy, depth and field size ✓
   E. SSD only

   TMR and TAR are indp of SSD

T3.   TAR is:
   A. equal to the backscatter factor (BSF) at d$_{max}$
   B. independent of SAD
   C. used in calculation of timer settings for rotational therapy
   (D.) all of the above ✓
   E. none of the above

   dose rate in tissue over

   TAR$_{dmax}$ = BSF

T4.   A single posterior spine field is treated at 130 cm SSD. Compared with treatment at 80 cm SSD the exit dose will be:
   (A.) greater
   B. smaller ✓
   C. the same

   ↑SSD ↑PDD

   as SSD↑ PDD↑ per Mayneord's f factor. and inv square law.

T5.   Choose the statement that is true:
   A. TAR increases as SSD increases *is dependent*
   B. BSF increases as beam energy increases above 1 MV *(only increases in low E range, then decreases as E↑)*
   (C.) PDD increases with increasing SSD ✓
   D. TMR cannot be measured for cobalt-60

T6-8.   Match the formula with the quantity:
          A. (dose rate at d_{max}) / (dose rate at depth) at SSD
          B. (dose rate at depth) / (dose rate at d_{max}) at SSD
          C. (dose rate at d_{max}) / (dose rate at depth) at SAD
          D. (dose rate at depth) / (dose rate at d_{max}) at (SAD)
          E. (dose rate at d_{max}) / (dose rate in air) at SAD

T6.   D   TMR   D

T7.   E   BSF   E   ⇒ is TAR @ dmax

T8.   B   PDD/100   B

T9.   E   TMR:
          A. stands for tumor-maximum ratio
          B. is the ratio of dose at d_{max} to dose at depth
          C. increases as SSD increases
          D. cannot be measured on a cobalt-60 unit
          E. none of the above ✓

T10. A   Back scatter factor: @ SAD
          A. is the TAR at d_{max} ✓
          B. increases as energy increases over 1 MV   — forward scatter
          C. is the PDD at d_{max}
          D. is the ratio of dose in air to dose in tissue
          E. all of the above

T11. C   All of the following are independent of SSD *except*:
          A. TMR
          B. TAR
          C. PDD ✓
          D. BSF

T12-15.   Match the parameter with its definition, according to the diagrams below.
          A. D5 / D4
          B. D5 / Dl
          C. D3 / D4
          D. D3 / D2
          E. D1 / D4

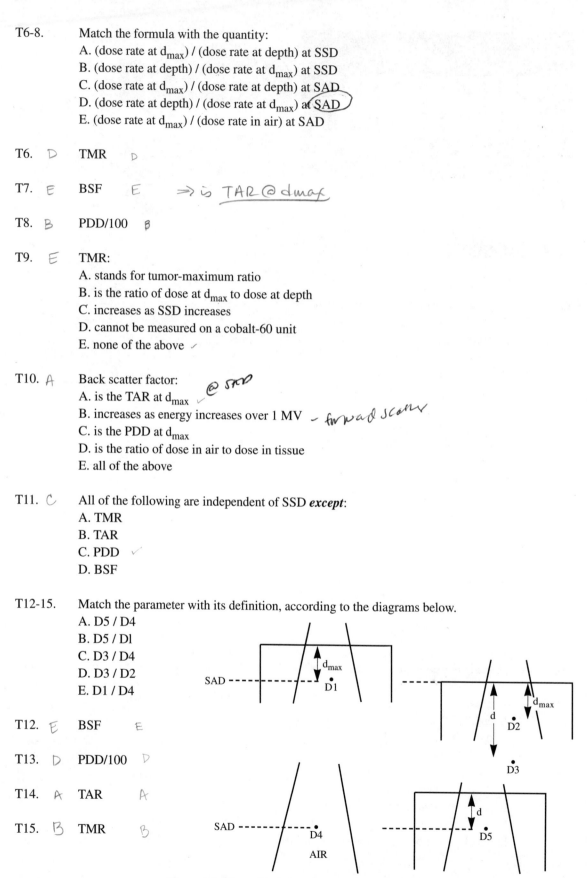

T12. E   BSF   E

T13. D   PDD/100   D

T14. A   TAR   A

T15. B   TMR   B

T16.   A patient is treated with cobalt-60 radiation in the manner shown below. Point A is 1 cm from the exit surface. The dose at point A is calculated using PDD tables. Relative to the actual value, the calculated dose at point A is:

A. more than 15% high
B. slightly high
C. correct
D. slightly low
E. more than 15% low

actual dose is pretty low

no BSF

80 SSD

20 cm

BF is lacking

@ 19 cm depth •A

T17-19.   Match the backscatter factor for a 10 × 10 cm field with the energy of the radiation:

A. 1.00 — *can not happen → implies dose in air = dose in tissue*
B. 1.02
C. 1.035
D. 1.15
E. 1.26   *1 mm Cu HVL = 1.5 50%*

BSF cannot have a value = 1.0 b/c that would mean no difference b/w DR dmax of tissue DR in air always > 1

T17.   E   Superficial, 2.5 mm Al HVL (1.26) *h. peak*

T18.   C   Cobalt-60   (1.035)   35%

T19.   B   10 MV x-rays   *very small for high energies. 1.02*

T20.   Which of the following statements is *false* about TMR?

A. it is equal to TAR/BSF ✓
B. it is approximately related to the percent depth dose by an inverse square factor ✓
C. it is the ratio of the dose at depth divided by the dose at d_{max}, both measured at the isocenter ✓
D. it is dependent on SSD
E. it increases with increasing field size

T21.   TAR at d_{max} can be calculated from the BSF by:

A. multiplying by the SAR
B. dividing BSF by the collimator output factor
C. applying inverse square correction to the BSF
D. no need to calculate; they are the same at d_{max}

T22.   For megavoltage photons, TMR has replaced TAR because:

A. TMR is independent of field size
B. TAR is difficult to measure at high energies ✓ → *b/c Buildup-cap for ↑E had to be so large that it started to act like mini phantom*
C. TMR is preferable to TAR for rotation calculations
D. TMR is independent of depth of maximum dose

T23.   Given a square and rectangle of the same area, which would you expect to have the greater percent depth dose for ^{60}Co?   *symmetry*

A. the square ✓
B. the rectangle
C. they have the same depth dose

scatter from corners of a rectangle have further to travel

T24. The TAR for a 10×10 cm field at 100 cm SAD for 4 MV photons is 0.8 at a depth of 7 cm. What change would you expect in the TAR by extending the SSD from 93 cm to 193 cm (200 cm from the source), for the same field size at SAD?

A. 5% increase in TAR
B. 10% increase
C. 15% increase
D. no change in TAR

does not depend on SAD *qualities of beam $\uparrow$ more head scatter less energy ramp dose out of beam so sweet to travel to beam*

T25. As photon energy increases, the tissue-maximum ratio at d_{max}:

A. increases
B. decreases
C. remains unchanged

TMR dependent on FS, energy, depth $\uparrow$ dose $2-3\%$

@ $d_{max} = 1.000$

T26. As photon energy increases, the % transmission through a 1 cm Lucite blocking tray:

A. increases
B. decreases
C. remains unchanged

T27. Which of the following is correct?

A. TAR = TMR × BSF
B. TAR = TMR / BSF
C. TMR = BSF / TAR
D. TMR = TAR × BSF
E. BSF = TAR × TMR

($\frac{d_{max}\ tissue}{air}$) × ($\frac{air}{tissue}$)   *TMR = $\frac{TAR}{BSF}$*

TAR = TMR (BSF)

T28. Percentage depth dose in photon beams:

1. increases with increasing SSD
2. increases with increasing field size
3. increases with increasing beam energy
4. decreases exponentially (not including inverse square effect and scattering) beyond d_{max}

A. 1 only
B. 1, 2, 3
C. 2, 4
D. 4 only
E. all are correct

inverse square law becomes less impt @ greater SSDs

PD2. Timer Settings, Monitor Units for SSD Set Up

T29. A patient is to be irradiated to a depth of 7 cm using a field size of 10×10 cm at 80 cm SSD. The PDD at this point is 65% and the dose rate at d_{max} is 100 cGy/min. The time to deliver 150 cGy is ___ min.

A. 0.65
B. 1.25
C. 1.5
D. 2.3
E. 2.7

$\frac{150}{(.65 \times 100)}$ $\frac{150\ cGy}{100\ cGy/min\ (.65)}$ = 2.3 min

T30. A patient is simulated with a 10×10 cm field at 80 cm SAD to deliver 200 cGy at 8 cm depth on a cobalt-60 unit. If the technique is now changed to 80 cm SSD, what values are required to calculate the timer setting?
1. Percent depth dose for a 10×10 cm field, 8 cm depth, 80 cm SSD
2. Dose rate in air at 80 cm for a 10×10 cm field
3. Dose rate in air at 80.5 cm for a 10×10 cm field
4. Dose rate at 80.5 cm in air for a 9×9 cm field
5. TAR for a 10×10 cm field at 8 cm depth
6. Backscatter factor for a 10×10 cm field
7. Backscatter factor for a 9×9 cm field
8. Percent depth dose for a 9×9 cm field, 8 cm depth, 80 cm SSD

A. 4, 7, 8
B. 3, 6, 1
C. 2, 6, 5
D. 4, 5
E. 2, 8

T31. A single direct field is set up at 80 cm SSD. The prescribed dose is 3000 cGy in 10 fractions at 5 cm depth. The dose rate at d_{max} is 100 cGy/min. The PDD at $d_{5cm} = 78\%$. The timer setting and dose at d_{max} are:
A. 3.00 mins and 385 cGy
B. 3.00 mins and 300 cGy
C. 3.85 mins and 300 cGy
D. 3.85 mins and 385 cGy

T32. Which of the following would be a correct formula to calculate the timer setting for the following? A single posterior spine field is treated at 80 SSD on a cobalt-60 unit; dose per fraction = 300 cGy at 3 cm depth; field size on skin = 6×12 cm (assume equivalent square = 8 cm).
A. $t = 30/$(Dose rate in air at 30 cm) $\times$ TAR (8×8, d_{3cm})
B. $t = 300/$(Dose rate at d_{max}, 80 cm SSD)
C. $t = 300/$(Dose rate in air at 80 cm) $\times$ PDD (8×8, d_{3cm})
D. $t = 300/$(Dose rate at d_{max} 80 cm SSD) $\times$ PDD (8×8, d_{3cm})
E. $t = 300/$(Dose rate at d_{max} at 80 cm SAD) $\times$ BSF (8×8)

T33. A single, direct spine field, 6×20 cm, is treated on a cobalt-60 unit at 80 cm SSD. The dose is prescribed at 4 cm depth. In order to calculate the timer setting in the most direct way, one would need all of the following data *except*:
A. relative output factor for 9×9 cm equivalent square
B. PDD (9×9 cm, d_{4cm})
C. TMR (9×9 cm, d_{4cm})
D. dose rate in tissue at d_{max}, 80 cm SSD for a 10×10 cm field
E. prescribed dose per fraction at $d = 4$ cm

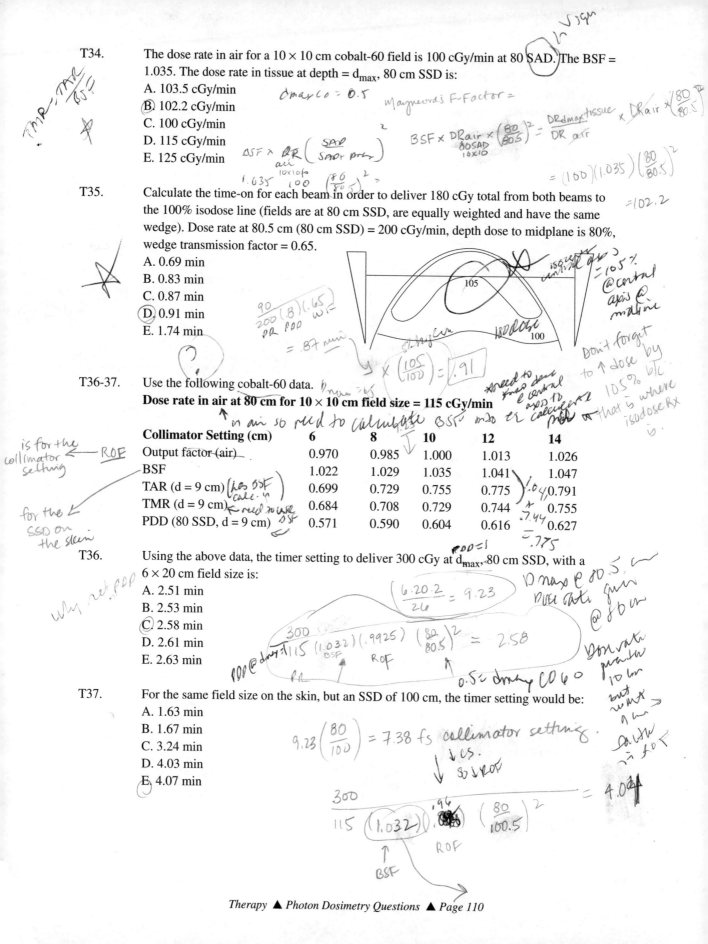

T34. The dose rate in air for a 10×10 cm cobalt-60 field is 100 cGy/min at 80 SAD. The BSF = 1.035. The dose rate in tissue at depth = d_{max}, 80 cm SSD is:

A. 103.5 cGy/min
B. 102.2 cGy/min
C. 100 cGy/min
D. 115 cGy/min
E. 125 cGy/min

T35. Calculate the time-on for each beam in order to deliver 180 cGy total from both beams to the 100% isodose line (fields are at 80 cm SSD, are equally weighted and have the same wedge). Dose rate at 80.5 cm (80 cm SSD) = 200 cGy/min, depth dose to midplane is 80%, wedge transmission factor = 0.65.

A. 0.69 min
B. 0.83 min
C. 0.87 min
D. 0.91 min
E. 1.74 min

T36-37. Use the following cobalt-60 data.
Dose rate in air at 80 cm for 10×10 cm field size = 115 cGy/min

Collimator Setting (cm)	6	8	10	12	14
Output factor (air)	0.970	0.985	1.000	1.013	1.026
BSF	1.022	1.029	1.035	1.041	1.047
TAR (d = 9 cm)	0.699	0.729	0.755	0.775	0.791
TMR (d = 9 cm)	0.684	0.708	0.729	0.744	0.755
PDD (80 SSD, d = 9 cm)	0.571	0.590	0.604	0.616	0.627

T36. Using the above data, the timer setting to deliver 300 cGy at d_{max}, 80 cm SSD, with a 6×20 cm field size is:

A. 2.51 min
B. 2.53 min
C. 2.58 min
D. 2.61 min
E. 2.63 min

T37. For the same field size on the skin, but an SSD of 100 cm, the timer setting would be:

A. 1.63 min
B. 1.67 min
C. 3.24 min
D. 4.03 min
E. 4.07 min

T38.　For parallel opposed 80 cm SAD fields, separation 18 cm, field size at midplane 12 × 12 cm, 200 cGy total dose per fraction, the timer setting per beam is:

A. 1.07 min
B. 1.11 min
C. 1.15 min
D. 1.39 min
E. 1.41 min

PD3. Timer Settings, Monitor Units for SAD Set Up

T39.　The dose rate in air at 80 cm is 120 cGy/min. The TAR for a 15 × 15 cm field at 12 cm depth is 0.686. A small corner block is used, and the tray factor is 0.96. The time to deliver 90 cGy is ___ min.

A. 0.54
B. 1.05
C. 1.09
D. 1.14
E. 1.26

T40.　To calculate the timer setting on a cobalt unit to deliver a dose of 200 cGy at 8 cm depth for a field size of 10 × 10 cm at the tumor, set up at 80 cm SAD, the values required are:

1. Percent depth dose for a 10 × 10 cm field size, 8 cm depth, 80 cm SSD
2. Dose rate in air at 80 cm for a 10 × 10 cm field size
3. Dose rate in air at 80.5 cm for a 10 × 10 cm field size
4. Dose rate at 80.5 cm in air for a 9 × 9 cm field size
5. TAR for a 10 × 10 cm field size at 8 cm depth
6. Backscatter factor for a 10 × 10 cm field size
7. Timer (shutter) error for the cobalt-60 unit
8. Backscatter factor for a 9 × 9 cm field size
9. Percent depth dose for a 9 × 9 cm field, 8 cm depth, 80 cm SSD

A. 3, 6, 1, 7
B. 4, 8, 9
C. 2, 1, 7
D. 4, 8, 1
E. 2, 5, 7

T41. A patient's pelvis is treated with parallel opposed 8 MV x-ray fields, set up at midplane, at 100 cm SAD. The prescribed dose is 4500 cGy in 5 weeks, 5 fractions per week. The AP thickness is 25 cm. Field size = 15 × 22 cm. Both fields are treated each day.

depth (cm)	PDD (100 cm SSD)	TMR
2	100	1.000
12	66.5	0.795
13	63.5	0.759
23	38.8	0.564

Output at 2 cm depth: at 100 cm SSD: 1.037 cGy/MU
at 100 cm SAD: 1.078 cGy/MU

The MU setting per beam is:
A. 134
B. 214
C. 268
D. 112
E. 107

T42. Calculate the timer setting for the following treatment: cobalt-60 unit, isocentric set up; collimator setting = 20 × 20 cm parallel opposed fields; total midplane dose per fraction is 200 cGy; patient thickness along axis = 22 cm. For 20 × 20 cm field size and 11 cm depth: PDD = 57.2%, TAR = 0.762. Output in air at 80 cm SAD = 125.5 cGy/min. Output at d_{max} in tissue at 80 cm SSD = 131.3 cGy/min. Timer error: add 0.02 mins.
A. 1.05 mins
B. 2.11 mins
C. 1.07 mins
D. 1.35 mins
E. 2.68 mins

T43. Timer settings have been calculated for a computer generated treatment plan with wedges. The fields are isocentric and the plan is for a cobalt-60 unit. The minimum information required to check the times by hand is:
1. Dose rate in air at the isocenter for each collimator setting
2. TAR tables
3. The relative contributions from each beam to the isocenter
4. The ratio of isodose at the isocenter to isodose level at which the dose is prescribed
5. Wedge transmission

A. 1, 2, 3, 4
B. 3, 4, 5
C. 1, 3, 4, 5
D. 1, 2, 3, 4, 5
E. 1, 2, 3

T44. A total dose of 200 cGy is to be delivered at a depth of 10 cm, at 100 cm SAD, from parallel opposed 15×15 cm 6 MV x-ray fields. Given the information below, the MU setting per beam would be:

TMR $(15 \times 15$ cm, $d_{10cm}) = 0.806$
Output in air, 100 cm SAD for 15×15 cm field size = 1.015 cGy/MU *← anytime a dose rate is expressed in air you need BSF* $\frac{1}{BSF} = 117$
BSF $(15 \times 15$ cm$) = 1.044$
PDD (100 cm SSD, 15×15 cm, $d_{10cm}) = 72.2\%$

$$\frac{100}{1.015(.806)(1.044)}$$

output   TMR   BSF

A. 123 MU
B. 125 MU
C. 136 MU
D. 117 MU
E. none of the above

T45. *≤ 1 reference field*
The dose rate for a 10×10 cm cobalt-60 field is 100 cGy/min in air, at 80 SAD. The dose rate (cGy/min) in air for a 30×30 cm field is about:
A. 101
B. 103
C. 107
D. 115
E. 125

$\frac{5 \times 5}{.96}$ *to* $\frac{30 \times 30}{1.07}$ *just have to memorize*

range of output factor for ^{60}Co *.95 — .107*

T46. A 6 MV linac is calibrated at the isocenter at depth d_{max} (D). To calculate the monitor units, MU = (dose at depth)/x, where x is equal to: *in tissue*
A. D × BSF × TAR
B. D × TAR *— if you use TAR need dose rate in air so* $\frac{D}{BSF} \times TAR = $ *dose rate in air* *D × TMR*   *see answer for good summary*
C. D × TMR
D. D × BSF × TMR
E. none of the above

PD4. Dose at d_2 From Dose at d_1 for SSD Set Up

T47. A patient is to be irradiated to a depth of 7 cm using a field size of 10×10 cm at 80 cm SSD. The PDD at this point is 65% and the dose rate at d_{max} is 100 cGy/min. If 150 cGy is delivered at d_{7cm}, the dose at d_{max} is ___ cGy.
A. 100
B. 142
C. 150
D. 154
E. 231

$\frac{150}{.65} = 231$

150 cGy = only .65 total dose

$TMR = \frac{TAR}{BSF}$

T48. A patient has previously been treated on a cobalt-60 unit to the supraclavicular region with a direct anterior field (SSD set up). Before treating the patient's spine, the dose to the cord from the supraclavicular field must be calculated. The data needed to do this is:

1. Given dose to supraclavicular field
2. Field size
3. Depth to cord
4. PDD tables
5. Timer error
6. Dose rate in tissue

A. 1, 2, 3, 4
B. 1, 2, 3, 4, 5
C. 2, 3, 4
D. 1, 2, 4, 5, 6
E. 3, 4, 6

T49. A 10×10 cm supraclavicular field is treated at 80 cm SSD on a cobalt-60 unit to a depth of 3 cm at 300 cGy per fraction. What is the dose/fraction at the cord if it is located at a depth of 7 cm? (Output 10×10 cm field size at 80 cm SSD = 200 cGy/min at depth = 0.5 cm.)

depth	% depth
0.5 cm	100%
3.0 cm	90%
5.0 cm	80%
7.0 cm	70%

A. 233 cGy
B. 267 cGy
C. 300 cGy
D. 338 cGy
E. 386 cGy

T50. A patient's separation is 20 cm. For parallel opposed fields treated at 100 cm SSD, what is the total dose to 2 cm, as a percentage of the total midplane dose?

depth	depth dose	TMR
2 cm	100%	1.00
10 cm	70%	0.80
18 cm	45%	0.60
20 cm	40%	0.55

A. 90%
B. 100%
C. 103.5%
D. 105%
E. 113%

T51. A 10×10 cm cobalt-60 beam is incident on an inclined surface, as shown in the diagram. PDD at $d_{6cm} = 0.739$; PDD at $d_{9cm} = 0.604$. If the dose at A is 100 cGy, the dose at B is approximately:

A. 76 cGy
B. 88 cGy
C. 114 cGy
D. 118 cGy
E. 132 cGy

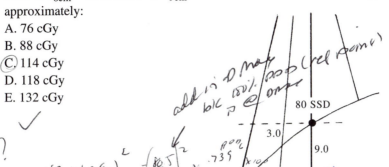

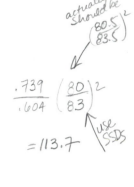

(handwritten annotations:) add in D_{max} bk 100%, pdd (rel point) P @ omg

80 SSD
3.0
9.0
6.0 .739   .604
100
B   A

$\left(\dfrac{SSD + d_{max}}{SSD + d_{max}}\right)^2 = \left(\dfrac{80.5}{83.5^2}\right)^2 \times \dfrac{.739}{.604} \times 100$

actually should be $\left(\dfrac{80.5}{83.5}\right)^2$

$\dfrac{.739}{.604}\left(\dfrac{80}{83}\right)^2 = 113.7$ use SSDs

$\dfrac{1.22}{.92} \quad \dfrac{.739}{.604} \times 100$

PD5. Dose at d_2 From Dose at d_1 for SAD Set Up

T52. A patient's pelvis is treated with parallel opposed 8 MV x-ray fields, set up at midplane, at 100 cm SAD. The prescribed dose is 4500 cGy in 5 weeks, 5 fractions per week. The AP thickness is 25 cm. Field size = 15×22 cm. Both fields are treated each day.

(handwritten:) 12.5 d   17.5   90/field

depth (cm)	PDD (100 cm SSD)	TMR
2	100.0	1.000
12	66.5	0.795
13	63.5	0.759
23	38.8	0.564

(handwritten: @ 12.5 { .777 ; $\left(\dfrac{.00}{89.5}\right)^2$; $\dfrac{138}{}$)

The total dose per fraction at 2 cm depth is:

A. 162 cGy
B. 175 cGy
C. 180 cGy
D. 198 cGy
E. 225 cGy

(handwritten:) you have to use $1/r^2$ correction for isocentric setup!

$D = 90 \left(\dfrac{100}{71}\right) \times \left(\dfrac{100}{89.5}\right)^2$

$$D(d_2) = D(d_1) \times \dfrac{TMR_2}{TMR_1} \times \left(\dfrac{SAD_1}{SAD_2}\right)^2$$

$+ \ 90\left(\dfrac{.56}{.77}\right) \times \left(\dfrac{100}{110.5}\right)^2$

T53. The attenuation of a photon beam is 4% per cm. If the dose at point P is 100 cGy, the dose at point Q (ignoring off center ratios and scatter differences) is:

A. 96 cGy
B. 100 cGy
C. 104 cGy
D. 92 cGy
E. 108 cGy

(handwritten:) $= (100)\ 1.04\left(\dfrac{100}{102}\right)^2$ use SADs

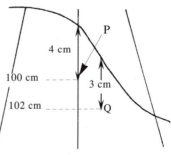

P
4 cm
100 cm
3 cm
102 cm
Q

$D_2 = 100 \times 1.04 \times \left(\dfrac{100}{102}\right)^2$

T54. A mantle is treated at 100 cm SAD, at d = 8 cm on the axis. If the dose is 4000 cGy at the axis, the total dose to the neck (ignoring off-axis factor) is:

	SSD (cm)		Separation (cm)
	AP	PA	
CAX	92	92	16
neck	93	97	10

depth (cm)	3	5	7	8
TMR	.987	.945	.901	.879

A. 3721
B. 3950
C. 4000
D. 4300
E. 4613

(handwritten) $\frac{.945}{.879} \times 4000 = 4300$

(handwritten) $\left(\frac{.945}{.879}\right)4000 = 4300$

(handwritten) a better way is to suppose midpoint to neat

PD6. Dose Variation Along Axis for Parallel Opposed Fields

(handwritten) and inverse sq corrections can out for midplane of neck

T55. Patient thicknesses of up to ___ cm can be treated on a cobalt-60 unit with parallel opposed 15 × 15 cm fields at 80 SSD if the total dose at d_{max} is not to exceed 110% of the midplane dose.
A. 8 cm
B. 14 cm
C. 20 cm
D. 24 cm
E. 28 cm

T56. A patient is treated with AP/PA isocentric fields to the mediastinum. The total midplane dose on the axis is 4500 cGy. The dose to the spinal cord will exceed 4500 cGy:
A. if the patient is treated on a cobalt-60 unit
B. because the cord is not a midplane structure
C. if the patient's AP thickness is less at the neck end of the field
D. all of the above

T57. If 3000 cGy is delivered at midplane (separation 20 cm) with parallel opposed 6 × 6 cm fields at 80 cm SSD, the maximum tissue dose is:

depth (cm)	0.5	10.0	19.5
PDD	100.0	52.5	24.1

A. 3000 cGy
B. 3300 cGy
C. 3546 cGy
D. 3784 cGy
E. 3194 cGy

(handwritten) $1500\left(\frac{100}{52.5}\right) + 1500\left(\frac{24.1}{52.5}\right) = 2857$

(handwritten) 3544

(handwritten) 100 is at D max by definition

T58.   If 4000 cGy is delivered at midplane to a patient's mediastinum, via parallel opposed fields, the lowest cord dose will result from treatment on a _____ unit, using _____ technique.
A. cobalt-60, isocentric
B. 8 MV x-ray, isocentric
C. 16 MV x-rays, isocentric
D. 16 MV x-ray, SSD   → maximize PDP
E. cobalt-60, SSD

[handwritten: A / highest Ng]

T59.   The total dose delivered at depth d_{max} from a pair of parallel opposed fields, expressed as a % of the total dose at midplane:
A. decreases as photon energy increases
B. decreases as field size increases
C. increases as patient thickness increases
D. is slightly less for an SSD set up than for an SAD set up
E. all of the above

[handwritten: $\dfrac{dose\ @\ dmax}{dose\ @\ midplane}$]

T60.   In a department with a cobalt-60 unit and a 10 MV linac, it is decided to treat parallel opposed isocentric pelvic fields on the cobalt-60 unit unless the total dose at d_{max} is more than 15% greater than the midplane dose. Patients with AP separations greater than ___ cm will be treated on the linac.
A. 10
B. 15
C. 20
D. 25
E. 30

T61-63.   The diagram shows the total dose at d_{max} (entrance + exit) as a % of the total dose at midplane, for 15×15 cm parallel opposed SSD fields. Match the curves with the appropriate beam energy.
A. cobalt-60
B. 20 MeV electrons
C. 6 MV x-rays
D. 10 MV x-rays
E. 25 MV x-rays

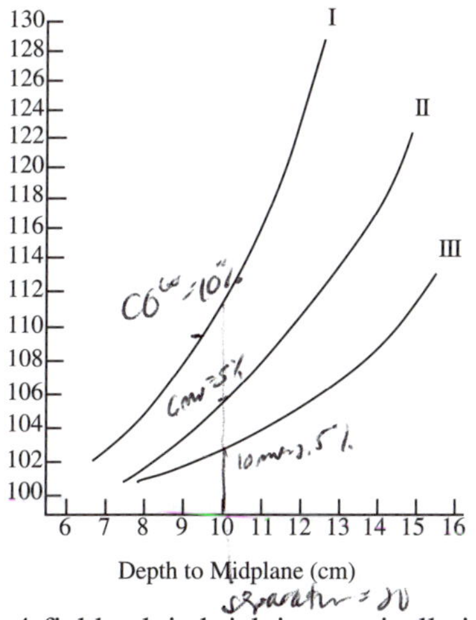

[handwritten X]

T61.   Curve I   A

T62.   Curve II   C

T63.   Curve III   D

[graph: y-axis 100–130, x-axis "Depth to Midplane (cm)" 6–16; curves labeled I, II, III; handwritten notes: Co⁶⁰ ~10%, Cru ~5%, 10 mv ~2.5%; separation = 20]

T64.   A possible disadvantage of treating a 4-field pelvic brick isocentrically instead of at SSD is:
A. increased set-up time   — need to move pt every field for SSD
B. field heights are different for AP and laterals
C. increased treatment time
D. increased dose to normal tissue   for SAD

(handwritten top) $\text{Mayneord's f Factor} = \left(\dfrac{SSD_2 + d_{max}}{SSD_1 + d_{max}}\right)^2 \times \left(\dfrac{SSD_1 + d}{SSD_2 + d}\right)^2$

PD7. Variations in PDD With SSD

T65. If the PDD of a beam at 10 cm depth is 75% at 100 cm SSD, the PDD at 10 cm depth for 80 cm SSD is: (d_{max} is 0.5 cm)
A. 70.5%
B. 71.9%
C. 75.0%
D. 78.5%
E. 80.0%

(handwritten) $.75\left(\dfrac{80.5}{100.5}\right)^2 \left(\dfrac{110}{90}\right)^2 = 71.9\%$
$\times 4 \qquad 1.49$
has to be a smaller %DD b/c of ↓ SSD

T66. The PDD at a new SSD can be calculated from the PDD at a standard SSD by multiplying by:
A. the f-factor, which is 0.957 for cobalt-60 *(handwritten: R to cGy conversion factor for Co-60)*
B. an energy dependent conversion factor
C. Mayneord's f-factor, which depends on SSD and depth only
D. the BSF for the new SSD
E. none of the above

(handwritten left: (exp → abs))

T67. The Mayneord f-factor which calculates the change in central axis depth dose with SSD includes corrections for:
A. penumbra
B. field size
C. scattered radiation
D. tissue absorption
E. inverse square law *(handwritten: only)*

T68. In order to obtain a large enough field to cover the leg, a patient is treated at 200 cm SSD, using 6 MV x-rays (d_{max} = 1.5 cm). The field projects to an equivalent square of 30×30 cm at 200 cm SSD. The PDD at 10 cm depth at 100 cm SSD for 30×30 cm equivalent square is 0.730. The PDD at the treatment SSD of 200 cm will be:
A. 0.730
B. 0.789
C. 0.759
D. 0.702
E. 0.675

(handwritten) $.73\left(\dfrac{201.5}{101.5}\right)^2 \left(\dfrac{110}{210}\right)^2 = .789$
$3.94 \qquad .27$
has to be a larger %DD b/c of increasing SSD

PD8. Equivalent Square

T69. The equivalent square of a 9×17 cm field size is:
A. 1.36
B. 6.84
C. 10.00
D. 11.76
E. 13.32

(handwritten) $\dfrac{a \times b \times 2}{a + b}$
$\dfrac{(9 \times 17) \times 2}{9 + 17} = 11.77$

T70-73.   The equivalent square of a rectangular field (answer A for true and B for false):

T70.   B   Has the same area as the rectangle.

T71.   B   Is approximately twice the area divided by the perimeter. $2\,area/perimeter$

T72.   A   Has the same PDD on the axis as the rectangular field. $\dfrac{2AB}{A+B}$

T73.   A   Has the same backscatter factor as the rectangular field.

T74.   For a field size of 15×22 cm, the side of the equivalent square field is:
A. 18.2 cm
B. 17.8 cm 17.8
C. 15.0 cm
D. 18.5 cm
E. 22.0 cm

T75.   Given a circle and rectangle of the same area, which would you expect to have the greater percent depth dose?
A. the circle
B. the rectangle
C. they have the same depth dose

PD9. Wedges

T76-79.   Using two wedge fields, a uniform dose distribution is usually obtained (answer A for true and B for false):

T76.   B   Only when the wedges are used at 90° to each other.

T77.   B   Only when a third open field is added.

T78.   A   When the wedge angle is approximately 90° minus half the hinge angle. $90 - \dfrac{\theta}{2}$

T79.   A   When the thick ends of the wedges are adjacent to each other.

T80.   The angle between the beam axes in a "wedged pair" is 60°. The most appropriate wedge angle would be:
A. 15°
B. 30°
C. 45°
D. 60° $90 - \dfrac{60}{2}$
    $\dfrac{180-60}{2}=60$

T81.   In the diagram below, the wedge angle is:
       A. not seen in the diagram, as it is the actual angle of the wedge itself
       B. B on the diagram
       C. C on the diagram
       D. D on the diagram
       E. E on the diagram

 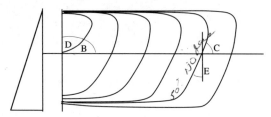

T82.   The treatment plan shown below has three 6 MV beams, each 8 × 8 cm, with 15° wedges on
       the lateral fields. Beams are weighted to deliver equal doses at the isocenter. The dose gradient
       across the tumor, from anterior to posterior, is 100-85%. How could the homogeneity of the
       dose distribution be improved?
       A. increase the wedge angle
       B. reverse the orientation of the wedges
       C. increase the weighting of the anterior field
       D. both B and C
       E. none of the above

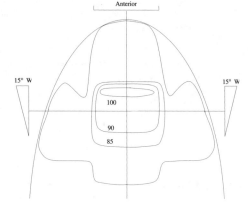

T83.   An effective wedge angle of 15° could be achieved by:
       A. use of a 30° wedge and an open beam for equal numbers of monitor units
       B. a universal wedge of 60°, combined with an open field by appropriate weighting on the
          treatment plan
       C. a combination of a 30° wedge and an open beam to deliver equal doses on the axis
       D. A, B and C
       E. B and C only

T84.   A single direct 20 × 20 cm field is set up at 80 cm SSD. The PDD at the treatment depth of 8
       cm is 68%. The dose rate at d_{max} is 100 cGy/min, and the prescribed dose is 150 cGy. If a
       wedge is placed in the beam, with wedge transmission factor (WTF) 0.75, the timer setting will
       be:
       A. 1.65 min
       B. 2.94 min
       C. 2.21 min
       D. 2.00 min

$$\frac{150}{100\,(.68)(.75)} = 2.94\ min$$

T85. A pituitary is treated with 2 lateral wedged fields and a coronal. Calculate the MU setting for the lateral fields, using the following data:
Prescribed dose = 4500 cGy in 25 fractions (at isocenter).
Beams are isocentric, and weighted to deliver equal doses at the isocenter.
Output in air at isocenter = 0.90 cGy/MU.
TAR = 0.782, and wedge factor = 0.58.

4500/25/3 = 60 cGy/beam/fraction

A. 147 MU ·
B. 85 MU
C. 49 MU
D. 162 MU
E. 255 MU

$$\frac{60}{.9\,(.782)(.58)} = 147$$

T86. The wedge transmission factor (WTF) is 0.59. The MU setting for an open beam is 150. The MU setting for the same dose, with the wedge in, would be:
A. 150 MU
B. 189 MU
C. 254 MU ·
D. 203 MU
E. 89 MU

$$\frac{150}{.59} = 254$$

T87. It is found that a 30° wedge has been inserted the wrong way for five out of a course of 30 treatments. If the radiation therapist wishes to correct for this error:
A. the timer setting for the next five treatments could be recalculated
B. the wedge should be reversed for the next five treatments
C. the wedge should be omitted for the next five treatments
D. A 60° wedge could be used (in the correct orientation) for the next five treatments, taking the new WTF into account ·
E. it is impossible to correct

PD10. Use of Wedges in Treatment Plans

T88. A treatment plan is performed for the set-up of a brain field and it is found that 60° wedges give the most uniform dose distribution. The reason why 60° wedges turned out to be better than 45° wedges is that:
A. the hinge angle requires 60° wedges ↲
B. the bone of the skull has a significant effect on the dose distribution, favoring a higher wedge angle
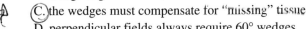
C. the wedges must compensate for "missing" tissue
D. perpendicular fields always require 60° wedges
E. the computer calculation was incorrect

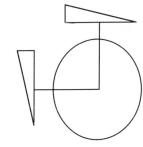

T89.   Select the field arrangement which would *not* give a uniform distribution.

A.   B.

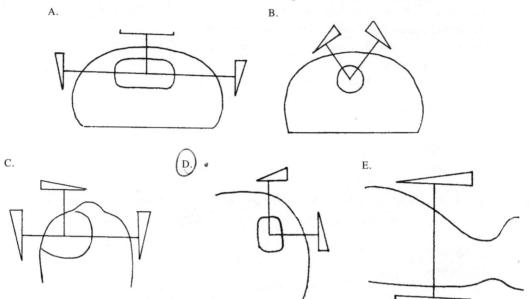

C.   D.   E.

T90.   The isodose distribution shown below is for a breast treatment given with a 15° wedge. Which of the following statements is true?
A. the isodose distribution could be improved if mixed with an open beam field
B. the isodose distribution could be improved if larger angle wedge is used ⌣
C. the isodose distribution could be improved if it were renormalized
D. the isodose distribution cannot be improved
E. the isodose distribution could be improved if just the open beam were used

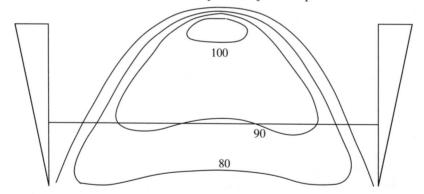

T91.   Wedges are used for all of the following *except*:
A. to increase homogeneity for breast tangent fields
B. to increase the anterior hot spot for bilateral larynx treatments ·
C. to increase homogeneity for orthogonal fields in partial brain treatments
D. as a compensator for a single upper spinal field with a sloping surface
E. to increase homogeneity in a 3 field plan (anterior and 2 opposing laterals)

T92. A wedged beam and an open beam of the same size are called up on a treatment planning system and given equal weight. The ratio of the wedged to open doses at d_{max} will be:
A. the wedge transmission factor (WTF) *- 100% × WTF*
B. 1.0 / WTF
C. 1.0 *or normalized to open beam (100%)*
D. A or C, depending on the treatment planning system
E. B or C, depending on the treatment planning system

1 if weighting factor is specified @ Dmax
but
if weighting is at tumor — then dose at Dmax is different

PD11. Rotation

T93. As a quick check of the MU setting for a 360° rotation plan you would need all of the following *except*:   *isocenter*
A. TAR (or TMR) tables
B. average AP/PA and lateral depth to isocenter
C. output in air (or at d_{max}) for collimator setting used
D. an isodose curve at standard SSD *·*
E. prescribed dose per fraction at the isocenter

T94. **Dose rate in air at 80 cm for 10×10 cm field size = 115 cGy/min**

FS (cm)	12	14
Output factor (air)	1.013	1.026
BSF	1.041	1.047
TAR (d = 17.5 cm)	0.500	0.520
PDD (d = 17.5 cm)	0.344	0.356

Using the beam data given above: A 100° lateral arc is treated on a cobalt-60 unit. The collimator setting is 12×12 cm. The average depth to the isocenter is 17.5 cm. The time needed to deliver 100 cGy to the isocenter is:
A. 1.65 min
B. 1.72 min
C. 1.74 min
D. 2.50 min
E. 2.53 min

$$\frac{100}{.5\,(1.013)(115)} = 1.72$$
↑
var op factor
½ Dose rate in air
just for 10×10 field
→ treatment field 12×12

T95. Which of the following is **not** a suitable treatment technique for the volume in the diagram?
A. two posterior oblique wedged fields
B. a posterior 180° arc, pinned at the center of the volume *→ isodose will fall off inside volume*
C. a posterior open field and wedged lateral fields (with thick ends posterior)
D. a posterior 180° arc, split into two 90° arcs with wedges (thick ends posterior)
E. none of the above

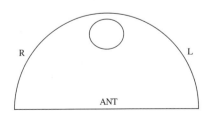

T96. A patient has an AP separation of 22 cm and a lateral separation of 32 cm. For a 360° rotational plan, the average TAR is approximately:

Depth (cm)	TAR
5	0.90
10	0.71
15	0.56
20	0.43
25	0.30
30	0.23

$$\frac{11 + 16}{2} = 13.5$$

A. 0.29
B. 0.38
C. 0.52
D. 0.60
E. 0.70

PD12. Surface Dose

photons surface dose ↓ as energy ↑

electrons surface dose ↑ as energy ↑

T97. Which of the following statements about surface dose in photon beams, as a percentage of dose at d_{max}, is *false*?

A. surface dose usually increases with the addition of a blocking tray
B. surface dose decreases as the SSD increases, for the same field size on the skin
C. surface dose increases as the photon energy increases
D. surface dose depends on the obliquity of the patient's skin surface
E. surface dose is primarily due to electrons generated in materials between the source and the patient

T98. Which of the following is *false*? The skin dose on a cobalt-60 teletherapy unit will increase when:

A. the SSD is decreased   b/c of collimator scatter
B. the field size is decreased
C. bolus is used
D. fields are treated at oblique incidence

T99. In treatment of Hodgkin's disease with a mantle field on a linac, patients may experience a skin reaction in the neck region. This could be due to all of the following *except*:

A. smaller thickness of tissue at neck than on the central beam axis
B. oblique incidence at sides of neck reduces skin sparing
C. use of a blocking tray increases skin dose over that in an open field
D. skin dose increases with field size, and mantles are large fields
E. linacs generally have high dose rates

T100. A patient's skin dose is reduced in a megavoltage photon beam by:

A. adding a 1 cm Lucite blocking tray
B. reducing the beam energy
C. increasing the field size
D. adding bolus
E. none of the above

PD13. Field Junctions

T101. When two AP/PA adjacent photon fields are needed to cover a treatment area, it is a good idea to have equal collimator settings for both fields because:

A. it is easier for the treatment set-up

B. the timer settings will be the same

C. a larger field size will diverge into a smaller field size and create hot and cold spots

D. the gap is easier to calculate

E. none of the above

T102. For a pair of adjacent fields (e.g., spinal fields) all of the following will help to improve dose uniformity in the field junction region *except*:

A. calculating a gap between the fields

B. angling the gantry for both beams

C. using a half-beam block on both fields

D. using a collimator angle for both beams

E. using a penumbra generator $\Rightarrow$ *a wedge that broadens penumbra so that matching is not as "critical"*

T103. A patient was treated to the thoracic spine using the cobalt technique outlined below. It is now necessary to treat the lumbar spine using a 6 MV x-ray beam. If the depth of the cord in the region of the junction is 6 cm, calculate the total gap which must be left on the skin surface to abut the fields at the cord.

Cobalt-60 technique
SSD: 80 cm
Field Size (cm): 20 (L) × 6 (W)
Prescription Depth: 4 cm

6 MV x-ray technique
SSD: 100 cm
Field Size: 30 (L) × 6 (W)
Prescription Depth: 8 cm

A. 1.50 cm

B. 1.70 cm

C. 3.40 cm

D. 1.65 cm

E. 2.00 cm

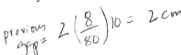

gap for "each" field:

$$g = d/SAD \times FS/2$$

depth of matching point. *@SAD*

$$g = \frac{6}{80}(10) + \frac{6}{100}(15) = 1.65\,cm$$

g_1 g_2

$$g = g_1 + g_2$$

T104. A patient is treated on a ^{60}Co unit with two 20 cm long spine fields, matched at 8 cm depth. The SSD for both fields is 80 cm. If the patient is now transferred to a linac and treated at 100 cm SSD using the same field sizes on the skin, how should the gap be changed so that the fields abut at the same depth as previously?

A. no change in gap

B. reduce gap by 1.0 cm

C. increase gap by 0.4 cm

D. none of the above

E. not enough information given

$$previous\ gap = 2\left(\frac{8}{80}\right)10 = 2\,cm$$

$$new\ gap = 2\left(\frac{8}{100}\right)10 = 1.6\,cm$$

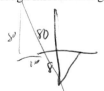

T105.   Two single direct spine fields are treated at 80 cm SSD. The collimator settings are 5×20 cm and 7×22 cm respectively. To match the fields at 5 cm depth a gap of _____ cm should be left on the skin.

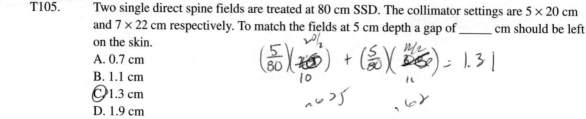

A. 0.7 cm
B. 1.1 cm
C. 1.3 cm
D. 1.9 cm
E. 2.6 cm

T106.   Adjacent pairs of parallel opposed AP/PA fields are to be matched at the patient's midplane. If the fields are treated isocentrically at 100 cm SAD, the collimator settings are 20 cm and 24 cm respectively, and the patient's AP thickness at the junction is 18 cm, the gap to be left on the skin is:

A. 1.0 cm
B. 1.98 cm
C. 4.0 cm
D. 2.5 cm
E. 5.0 cm

$$\frac{9}{100}\left(10\right) + \frac{9}{100}\left(12\right) = 1.98$$

T107.   The formula used to calculate the gap on the skin between adjacent fields, matched at depth, relies on the fact that:

A. the projection of the edge of the light field follows the 50% decrement line of the radiation field
B. both beams must be treated with the same energy x-ray beam
C. the angle of divergence of adjacent beams is the same
D. both fields must be treated simultaneously
E. the depth at the junction is less than 10 cm

T108.   Calculate the gap on the skin surface between the two following fields:

Field size (cm):	20 (L) × 10.5 (W)	30 (L) × 6 (W)
Depth (cm):	10	10
SAD (cm):	100	80

The depth to the matchpoint in the plane of the junction is 12 cm (note that the field dimensions are in cm).

A. 3.45 cm
B. 2.87 cm
C. 5.75 cm
D. 1.73 cm
E. 6.9 cm

$$10\left(\frac{12}{100}\right) + 15\left(\frac{12}{80}\right) = 3.45$$

T109.   A 20 cm length field was treated AP/PA and set-up at 100 cm SSD. A 25 cm length field (AP/PA) needs to be treated adjacent to the 20 cm field. What SSD for the 25 cm field could be used so that the divergence matches the 20 cm, 100 SSD field?

A. 80 cm SSD
B. 90 cm SSD
C. 100 cm SSD
D. 125 cm SSD
E. not enough information given

$$\frac{10}{100} = \frac{12.5}{x} \qquad gap?$$

$$x = 125 \text{ cm}$$

note: collimator setting on skin to be same.

T110. All of the following are acceptable methods of removing beam divergence at the chest wall border of tangentially opposed breast fields, except:
A. beam splitter
B. asymmetric collimators
C. gantry rotation so that beams are not exactly opposed ✓ } removes divergence
D. collimator rotation ?

T111. In the treatment of medulloblastoma the lateral brain fields can be given a collimator rotation to align them with the diverging spine field. If the spine field length is 25 cm at 100 cm SSD, the collimator rotation will be:
[data: tan 14° = 0.25; tan 7° = 0.125; tan 25° = 0.47; tan 2.5° = 0.044]
A. 14°
B. 7°
C. 25°
D. 2.5°
E. none of the above

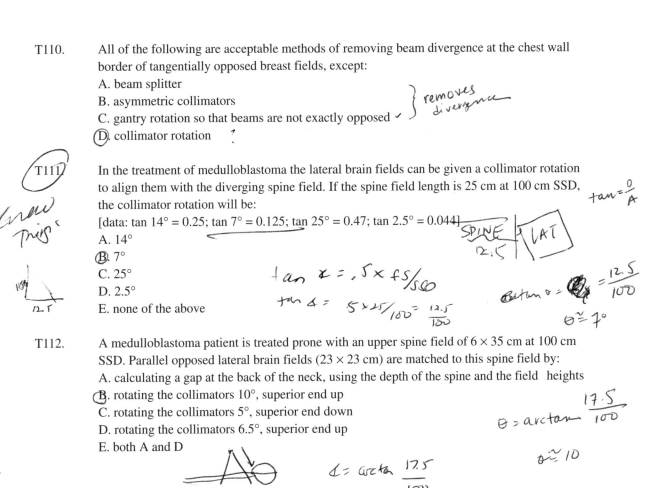

T112. A medulloblastoma patient is treated prone with an upper spine field of 6×35 cm at 100 cm SSD. Parallel opposed lateral brain fields (23×23 cm) are matched to this spine field by:
A. calculating a gap at the back of the neck, using the depth of the spine and the field heights
B. rotating the collimators 10°, superior end up
C. rotating the collimators 5°, superior end down
D. rotating the collimators 6.5°, superior end up
E. both A and D

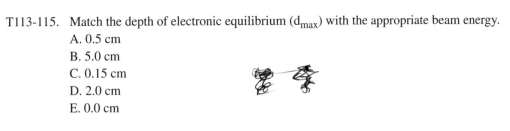

PD14. Depth Dose Variation With Energy

T113-115. Match the depth of electronic equilibrium (d_{max}) with the appropriate beam energy.
A. 0.5 cm
B. 5.0 cm
C. 0.15 cm
D. 2.0 cm
E. 0.0 cm

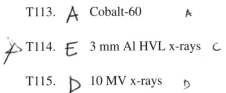

T113. A   Cobalt-60   A

T114. E   3 mm Al HVL x-rays   C

T115. D   10 MV x-rays   D

T116.   The depth of d_{max} is:
         A. dependent on photon energy
         B. dependent on field size
         C. independent of SSD for the same field size — *(handwritten: PDD ∞ S²)*
         D. all of the above ✓
         E. none of the above

T117-120. Match the approximate depth of 50% central axis depth dose with the appropriate beam.
           A. 2 cm
           B. 5 cm
           C. 8 cm
           D. 12 cm
           E. 15 cm

T117. **B**   12 MeV electrons, 100 cm SSD   *(handwritten: 5 cm   range = D/2 = 6   50 50% will be 6)*

T118. **D**   Cobalt-60 teletherapy, 80 cm SSD   *(handwritten: 2 cm (1 kV 4 mv))*

T119. **A**   120 kVp, 15 cm SSD, 2.0 mm HVL   *(handwritten: 2 cm   superficial)*

T120. **E**   6 MV linear accelerator, 100 cm SSD   *(handwritten: 15 cm @ 50% isodose (@ 10cm = 67%))*

T121-128. Match the photon beam with the description:
           A. cobalt-60, 80 cm SSD
           B. 4 MV x-rays, 100 cm SSD
           C. 10 MV x-rays, 100 cm SSD
           D. 25 MV x-rays, 100 cm SSD
           E. 2 mm Al HVL x-rays, 20 cm SSD

T121. **A**   Has a similar depth dose to a 14 MeV neutron beam.   *(handwritten: c)*

T122. **A**   Is not created by accelerating electrons.   *(handwritten: A)*

T123. **C** D_{max} is at 2.0 cm depth.   *(handwritten: c)*

T124. **A**   For a 10×10 cm field, the PDD at d_{10cm} is 56%.

T125. **B**   For a 10×10 cm field, the PDD at d_{10cm} is 63%.

T126. **E**   Has the greatest BSF.   *(handwritten: E ⟩ superficial o/...)*

T127. **E** D_{max} is at zero depth.   *(handwritten: E)*

T128. **D**   Would result in the lowest integral dose, when treating a 4-field pelvic brick.   *(handwritten: D   highest MV)*

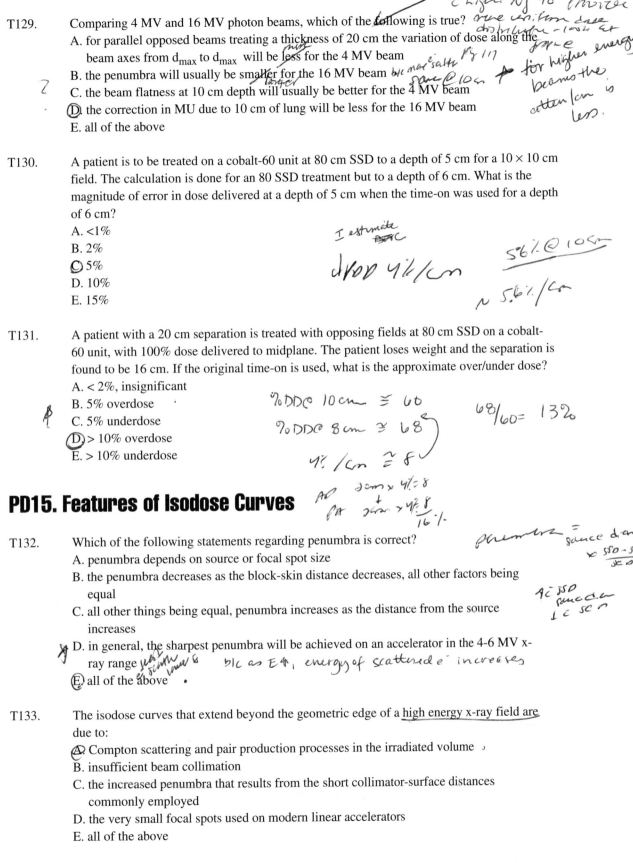

T129.	Comparing 4 MV and 16 MV photon beams, which of the following is true? *we treat layer pilloe / E lighter NJ to Conize / ore uniform dose / distrilution - look at / for higher energy*
	A. for parallel opposed beams treating a thickness of 20 cm the variation of dose along the *more* beam axes from d_{max} to d_{max} will be less for the 4 MV beam *P 117*
	B. the penumbra will usually be smaller for the 16 MV beam *b/c more sstts* *same @ 10cm* *for higher energy beams the*
	C. the beam flatness at 10 cm depth will usually be better for the 4 MV beam *larger* *otten /cm is*
	D. the correction in MU due to 10 cm of lung will be less for the 16 MV beam *less.*
	E. all of the above

	2

T130.	A patient is to be treated on a cobalt-60 unit at 80 cm SSD to a depth of 5 cm for a 10×10 cm field. The calculation is done for an 80 SSD treatment but to a depth of 6 cm. What is the magnitude of error in dose delivered at a depth of 5 cm when the time-on was used for a depth of 6 cm?
	A. <1%
	B. 2%
	C. 5% *I estimate BSC* *5.6% @ 10cm*
	D. 10% *drop 4%/cm* *~5.6%/cm*
	E. 15%

T131.	A patient with a 20 cm separation is treated with opposing fields at 80 cm SSD on a cobalt-60 unit, with 100% dose delivered to midplane. The patient loses weight and the separation is found to be 16 cm. If the original time-on is used, what is the approximate over/under dose?
	A. < 2%, insignificant
	B. 5% overdose *%DDC 10cm ≅ 60* *68/60 = 13%*
	C. 5% underdose *%DDC 8cm ≅ 68*
	D. > 10% overdose *4%/cm ≅ 8*
	E. > 10% underdose

	A
	AP 2cm × 4% = 8
	PA 2cm × 4% = 8
	16%

PD15. Features of Isodose Curves

T132.	Which of the following statements regarding penumbra is correct? *penumbra = source diameter × SSD-SCD / SCD*
	A. penumbra depends on source or focal spot size
	B. the penumbra decreases as the block-skin distance decreases, all other factors being equal *↑ SSD penumbra ↑ SCD*
	C. all other things being equal, penumbra increases as the distance from the source increases
	D. in general, the sharpest penumbra will be achieved on an accelerator in the 4-6 MV x-ray range *less scatter lower E* *b/c as E ↑, energy of scattered e⁻ increases*
	E. all of the above

T133.	The isodose curves that extend beyond the geometric edge of a high energy x-ray field are due to:
	A. Compton scattering and pair production processes in the irradiated volume
	B. insufficient beam collimation
	C. the increased penumbra that results from the short collimator-surface distances commonly employed
	D. the very small focal spots used on modern linear accelerators
	E. all of the above

T134.   Which of the following is true regarding geometric penumbra?
A. increases as source diameter increases
B. decreases as SSD increases
C. is independent of source-collimator distance
D. all of the above

T135.   The principal reason that the penumbra of a 4 MV linac is better than the penumbra of a cobalt-60 teletherapy unit is because for the linac:
A. output is higher
B. average energy is higher
C. SSD is longer
D. target or source size is smaller
E. flattening filter improves the edges

T136.   The penumbra of an x-ray beam will increase when:
A. the SSD is decreased
B. the source diameter is decreased
C. the depth is decreased
D. the source-collimator distance is decreased
E. the field width is decreased

T137.   On a cobalt-60 unit, instead of placing blocks on the usual shadow tray at 45 cm from the source, blocks are placed on a table whose surface is 10 cm above the patient's skin (at 70 cm). This is done to:
A. enable the use of smaller blocks than usual
B. minimize the scatter from the blocks
C. minimize the penumbral edges of the blocks
D. enable one less HVL of lead to be used.

T138.   The "horns" of an x-ray beam refer to the cross beam profile:
A. before the beam is flattened
B. for a wedged field
C. for a large field at d_{max}
D. when the beam is asymmetric
E. at 10 cm depth in water

T139.   The penumbra of a cobalt-60 beam increases with depth due to:
1. geometrical effects due to source diameter
2. inverse square effects
3. photon scatter, which increases with depth
4. collimator design

A. 1, 2, 3
B. 1, 3
C. 2, 4
D. 1 only
E. all of the above

PD16. Integral Dose

[handwritten: = energy deposited → mass of tissue (gm)]
[handwritten: = energy deposited, dose (cGy) × mass of tissue (gm)]
[handwritten: J/kg × .001kg = J = energy]

T140-143.   Integral dose (answer A for true and B for false):

T140.   A Can be measured in <u>megagram</u> cGys.   *[handwritten: A]*

T141.   A Is a measure of the <u>total energy</u> deposited in the patient.   *[handwritten: A]*

T142.   B Should be maximized in a treatment plan.   *[handwritten: should be minimized.]*

T143.   B Increases with increasing photon energy for a 4 field pelvic brick. ✓

T144.   Both treatment techniques shown below deliver a homogenous dose to the treatment volume,
             ± 8%. The <u>highest integral dose</u> would be delivered from:
             A. technique I, cobalt-60
             B. technique I, 8 MV x-rays
             Ⓒ technique I, cobalt-60
             D. technique II, 8 MV x-rays
             E. all would give equal integral doses

[handwritten: larger deposit, lower energy]

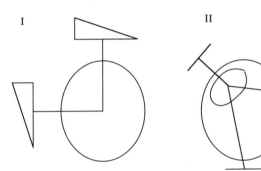

T145.   If a superficial tumor is treated with a single direct x-ray beam, which of the following will
             decrease the integral dose?
             A. increasing the prescription depth
             B. increasing field size
             C. increasing patient thickness
             Ⓓ decreasing beam energy
             E. using bolus

PD17. Variation in Dose Rate With Distance

T146. During routine quality assurance checks, the optical distance indicator (ODI) is found to read 100 cm when the distance is actually 98 cm. If uncorrected, the effect on treatments set up at 100 cm would be:

A. an overdose of 2%

B. an overdose of 4%

C. an overdose of 1%

D. an underdose of 2%

E. an underdose of 4%

$$\left(\frac{100}{98}\right)^2 = 1.04$$

T147. The dose rate in air for a cobalt-60 source is 100 cGy/min at 80 cm SAD. The dose rate in air for the same collimator setting at 200 cm is:

A. 50 cGy/min

B. 16 cGy/min

C. 250 cGy/min

D. 40 cGy/min

E. cannot be calculated without additional information

$$100 \times \left(\frac{80}{200}\right)^2 = 16$$

T148. The dose rate at a patient's midplane is found to be 250 cGy/min at 100 cm SAD. A protocol requires that the dose rate be no more than 100 cGy/min. The patient must therefore be treated at _____ cm to midplane.

A. 40

B. 250

C. 293

D. 158

E. 160

Dose rate × Dist² = Dose rate × Dist²

$$250 \times (100)^2 = 100 \,(x)^2$$

$$100 = 250\left(\frac{100}{x}\right)^2 \qquad x = 158$$

T149. A treatment plan for a patient treated with a mantle field was done assuming the patient was to be treated at 120 cm SAD. 207 monitor units (MU) were required to deliver the prescribed dose of 100 cGy to midplane (AP). At first set-up, it was decided to treat the patient at 140 cm SAD on the same machine. To a close approximation, how many monitor units are needed to treat the patient at the new distance?

A. 152 MU

B. 177 MU

C. 207 MU

D. 242 MU

E. 282 MU

$$(140)^2 \times MU = (120)^2 (207)$$

$$\left(\frac{120}{140}\right)^2 = .73 \qquad \frac{207}{.73} = .283 \qquad MU = 152 \text{ equal to } 207$$

$$\left(\frac{207}{152}\right)(207) = 282.$$

T150. A superficial x-ray unit has an SSD of 20 cm. An air gap of 2 cm, if no correction is made, would cause the dose at d_{max} to be:

A. 2% low

B. 4% low

C. 9% low

D. 17% low

E. 9% high

$$\left(\frac{20}{22}\right)^2 = 17\%$$

PD18. Bolus, Compensators, Beam Spoilers

T151.   Tissue compensating filters:
      A. have advantages over the use of bolus in high energy radiation therapy ✓
      B. must be made of material having the same density as tissue
      C. should be placed less than 15 cm from the patient's skin
      D. eliminate the skin sparing effect
      E. are not needed when dose computations are done with a computer

T152.   A patient with Hodgkin's disease is being treated with 10 MV x-rays. To increase the dose
      to the superficial nodes, a beam spoiler (1 cm thick Lucite board positioned above the patient's
      surface) is used. Which of the following statements regarding the effects of the spoiler is true?
      A. as the distance of the spoiler from the patient's surface increases, the surface dose will
         decrease • *b/c e⁻ scatter out of beam before reaching pt*
      B. as the thickness of the spoiler is increased from 1 to 2.5 cm, the surface dose will
         decrease
      C. the spoiler has exactly the same effect on the surface dose as bolus placed on the skin *does not raise dmax on skin surface*
      D. a beam spoiler is generally necessary for all Hodgkin's patients treated with energies of
         10 MV or less
      E. all of the above are true

T153.   Tissue compensating filters:
      A. enable uniform doses to be delivered but sacrifice skin sparing
      B. preserve the skin sparing but sacrifice uniform dose distribution
      C. are a substitute for bolus and preserve skin sparing
      D. eliminate penumbra effects

T154-156. Match:
      A. surface dose increases
      B. surface dose decreases
      C. no change in surface dose

T154. A Electron energy increases A

T155. B Photon energy increases B

T156. A Photon field size increases C   *↑ scatter from edges from e⁻*

T157.   Skin sparing is reduced by:
      A. use of any thickness of bolus
      B. treating through a plaster cast
      C. use of a beam spoiler
      D. treating in tissue under a skin fold (for example, breast treatment)
      E. all of the above •

T158.   The purpose of a "beam spoiler" is to:
         A. reduce the energy of a photon beam
         B. reduce the depth of penetration of an electron beam
         (C.) increase dose in the build-up region of a photon beam
         D. filter out scattered electrons from a photon beam, to reduce skin dose

PD19. Isodose Distributions

T159-163. Which of the following isodose patterns is appropriate for the field configurations and wedges
           shown? (answer A for correct, B for incorrect)

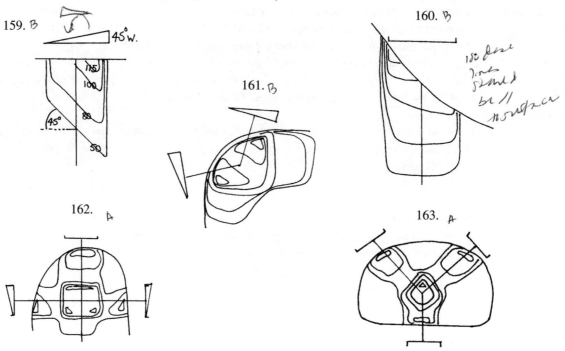

159. B   45° w.
          175
          100
          45°   80
                 50

160. B   iso dose
          ?ont
          should
          be //
          surface

161. B

162. A

163. A

T164-168. Match the treatment modality with its corresponding isodose curves.

T164.   22 MV x-rays   D

T165.   Cesium insertion E

T166.   Electron therapy A
          & 85% дпрас sme

T167.   Cobalt therapy B

T168.   200 kV x-rays C

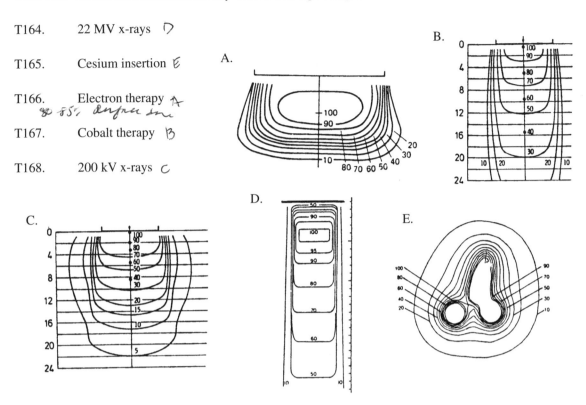

T169-172. Match the isodose distributions with the type of plan:

T169.   75 mg-Ra eq. cesium-137 intracavitary source C
T170.   Posterior 120° arc rotation, cobalt-60 A
T171.   45° wedged pair of fields, cobalt-60 B
T172.   360° rotation, cobalt-60   D

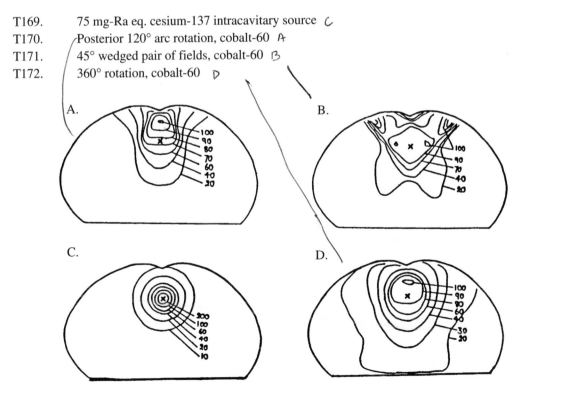

T173.   In the diagram of a breast plan (below), if the plan were re-normalized to the 95% level, then prescribed at 200 cGy per fraction to the new 100% line:

A. the variation of dose across the breast tissue would increase by 5%
B. the variation of dose across the breast tissue would increase by 10%
C. the maximum tissue dose would be 221 cGy per fraction
D. the maximum tissue dose would be 211 cGy per fraction
E. the hot spots on the plan would be 115%

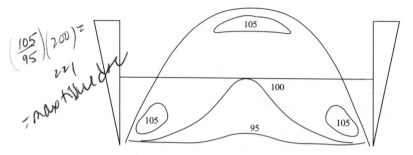

T174-178. In each of the following 5 plans, indicate the point of maximum dose. The plans are all for an 80 cm SSD cobalt-60 unit. * denotes the isocenter.

174.   175.

176.

177.   178.

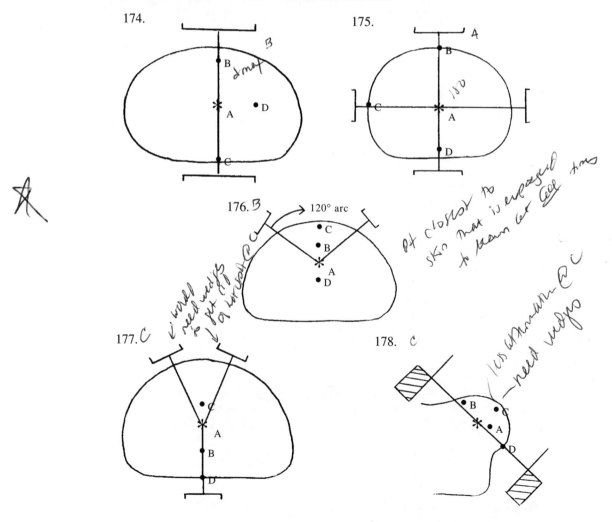

PD20. Superficial X-Rays

T179.   In order to increase the PDD at 0.5 cm for 100 kVp x-rays, the single most important step
         would be to:
         A. decrease the SSD from 20 cm to l5 cm
         B. increase the HVL from 1 mm to 3 mm Al
         C. increase the SSD from 15 cm to 20 cm   → would neither more it d was greater
         D. use only the inherent filtration of the x-ray tube   Maynard F= $\left(\frac{SSD_{i}+d_{max}}{SSD_{1}+d_{max}}\right)^{2}$ $\frac{2}{\left(\frac{SSD_{1}+d}{SSD_{2}+d}\right)}$
         E. increase the field size from 50 cm^2 to 100 cm^2

A

T180.   A patient is to be treated on a superficial x-ray unit with a 6 cm diameter cone in contact
         with a 4 cm diameter lead cutout. The data required to calculate the timer setting to deliver a
         given dose of 300 cGy are: dose @ surface
         1. PDD tables.   so don't need PDD
         2. The BSF for a 4 cm diameter field.
         3. The output in air for the 6 cm cone.
         4. The BSF for a 6 cm diameter field.
         A. 1 and 4
         B. 1, 2, 3, 4
         C. 1, 2, 3
         D. 1, 3, 4
         E. 2 and 3

T181.   Which of the following is **not** required to calculate the timer setting to deliver a prescribed
         given dose on a superficial x-ray machine?
         A. knowledge of which filter is to be used
         B. exposure rate in air for the applicator used   b/c dose specified @ surface
         C. BSF for the field size on the skin   for sup. x-rays
         D. PDD tables for the appropriate HVL and TSD
         E. measurement of any air gap between the applicator and the patient's skin

T182.   To select the appropriate PDD table from BJR Supplement 17 for a superficial x-ray unit,
         the two factors required are:
         A. field size and kVp
         B. kVp and SSD   SDD does affect the
         C. HVL and SSD   %DD in superficial
         D. filtration and SSD   machines
         E. filtration and field size

T183.   Two superficial x-ray therapy units both have 3 mm Al HVL. This means that one can
         assume:
         A. they have the same kVp
         B. they have the same kVp and mA
         C. if operated at the same SSD, they will have approximately the same percent depth dose
         D. they are both single phase, half-wave rectified
         E. they have the same inherent + added filtration

T184. A 1 mm Al filter is inadvertently used instead of a 2 mm filter on a superficial x-ray unit. The effect will be:
1. increased dose rate
2. increased PDD — ↓ Nᵧ c̄ more low Nᵧ rays in beam (not filtered)
3. decreased HVL
4. harder beam
A. 1, 2, 3
B. 1, 3
C. 2, 4
D. 4 only
E. all of the above

PD21. CT in Treatment Planning

T185. Which of the following is **not true** for CT images of the torso used directly for computerized treatment planning?
A. the patient must be scanned in the treatment position
B. a flat insert is required for the CT table
C. the CT image is a gray scale representation of the linear attenuation coefficient of each pixel
D. CT numbers must be converted into electron densities before pixel by pixel inhomogeneity corrections can be made
E. triangulation points or surface marks are unnecessary since the isocenter can be related to internal organs

T186. For accurate megavoltage treatment planning using CT images, which of the following is/are necessary?
1. a flat table insert
2. markers, such as plastic catheters, over skin marks on the patient (e.g. triangulation points)
3. patient in the treatment position (supine vs. prone, etc.)
4. treatment planning system with pixel-by-pixel heterogeneity correction
A. 1, 2, 3
B. 1, 3
C. 2, 4
D. 4 only
E. all of the above

T187. The Hounsfield number for water is:
A. 1000
B. 500
C. 0
D. -500
E. -1000

T188.   Which of the statements below is true concerning post-mastectomy treatments of the chest wall and supraclavicular areas?
A. these patients should always be treated with cobalt-60 as a first choice   ✗
B. electrons of the same energy can usually be used for both the chest wall and supraclavicular fields   ✗
C. electrons will always deliver less dose to the lungs than tangential photon beams   ✗
(D) lung corrections should be made only when a CT scan is available

PD22. Inhomogeneity Corrections

T189.   Part of a patient's treatment for cancer of the esophagus consists of the lateral fields described below. The ratio of treatment times calculated with and without inhomogeneity corrections would be: technique: 6 MV x-rays, 100 cm SAD, isocentric; field size: 20×6 cm at isocenter; lateral separation: 20 cm.

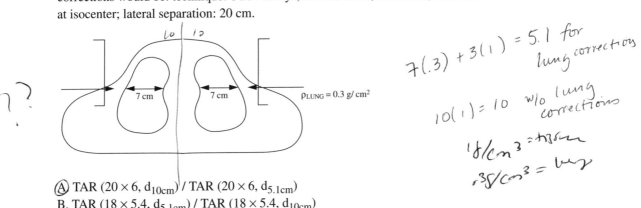

$7(.3) + 3(1) = 5.1$ for lung correction

$10(1) = 10$ w/o lung corrections

$1 g/cm^3 = tissue$
$.3 g/cm^3 = lung$

(A) TAR $(20 \times 6, d_{10cm})$ / TAR $(20 \times 6, d_{5.1cm})$
B. TAR $(18 \times 5.4, d_{5.1cm})$ / TAR $(18 \times 5.4, d_{10cm})$
C. TAR $(20 \times 6, d_{7cm})$ / TAR $(20 \times 6, d_{10cm})$
D. PDD $(100$ SSD, $20 \times 6, d_{5.1cm})$ / PDD $(100$ SSD, $20 \times 6, d_{10cm})$
E. PDD $(100$ SSD, $18 \times 5.4, d_{5.1cm})$ / PDD $(100$ SSD, $18 \times 5.4, d_{10cm})$

T190.   In the treatment of head and neck cancer, dosimetry corrections for the presence of bone and sinuses are most important for:
A. cobalt-60 γ-rays
B. 4 MV x-rays
C. 6 MV x-rays
(D) 14 MeV electrons . b/c more scatter
E. all of the above are equally important

T191-193. Regarding inhomogeneity corrections (answer A for true and B for false):

T191.   B The corrections are more important in a 10 MV photon beam than in a cobalt-60 beam. B
         ↑E ↓% correction

T192.   A 10 cm of lung in a cobalt-60 beam will increase the dose beyond the lung by about 35%.
         3.5%   electrons >

T193.   A In a cobalt-60 beam the correction for dense bone is about the same % per cm as for lung, but in the opposite direction. B

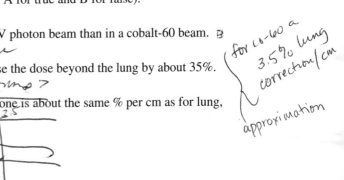

(for Co-60 a 3.5% lung correction/cm approximation)

T194. An isocentric 8 × 8 cm oblique beam has a total depth of 15 cm, and passes through 10 cm lung. Which of the following is *false*?

A. the % lung correction will be greater for cobalt-60 photons than for 8 MV x-rays

B. the lung correction for cobalt-60 is about 35%

C. calculations for time-on which ignore lung correction will deliver less dose to the tumor than prescribed *Mult*

D. using lung density to calculate equivalent path length to the isocenter is an approximation which overestimates dose just beyond the lung *(b/c ↓ scatter from lung)*

E. the density of an individual patient's lung can be derived from CT numbers in the region of the lung

PD23. Irregular Field Calculations

T195. Consider the two radiation portals shown below which have equal dimensions, SSDs and energies. The shaded regions represent beam shaping blocks, 5 HVL thick, of equal area. Which of the following is true?

↓ scatter to pt 2

Field #1 has a greater equiv. square.

A. the dose at point '1' is greater than the dose at point '2'

B. both fields have the same equivalent square

C. field '2' has a larger equivalent square than field '1'

D. A and C are correct

E. none of the above are true

T196. A patient is to be treated with cobalt-60 using an isocentric set-up with blocks and a collimator setting of 10 × 10 cm. The treatment planning computer performs a Clarkson calculation for the irregular field and yields an equivalent square field of 7 × 7 cm. The appropriate formula for computing the dose rate at isocenter is:

A. output in air (10 × 10 cm) × TAR (10 × 10 cm)

B. output in air (10 × 10 cm) × TAR (7 × 7 cm) ·

C. output in air (7 × 7 cm) × TAR (10 × 10 cm)

D. output in air (7 × 7 cm) × TAR (7 × 7 cm)

E. output in air (8.5 × 8.5 cm) × TAR (8.5 × 8.5 cm)

T197. The fraction of dose due to scatter is greatest for which of the following cobalt-60 fields?

A. 5 × 5 cm field at d_{max}

B. 10 × 10 cm field at d_{max} ·

C. 5 × 5 cm field at 10 cm depth

D. 10 × 10 cm field at 10 cm depth

E. all of the above have an equal fraction of scatter

↑ scatter ō depth & fs

T198.   The midline dose under a beam block is most readily calculated by using:

A. the Mayneord f-factor
B. narrow-beam attenuation coefficients
C. the assumption that the block behaves as a negative source
D. the Clarkson method *(dividing up into primary and scatter components)*
E. the inverse-square law modified by the area of the block

T199.   In the calculation of the timer setting for a mantle field, using the Clarkson method, TAR_0 represents:

A. the scatter component of the dose on the central beam axis
B. the primary component of the dose on the central beam axis ←
C. the backscatter factor for the blocked field
D. the TAR for the equivalent square of the blocked field
E. the tissue-air ratio for the open, unblocked field

↑ extrapolating to Ø field radius.

$TAR_{0ver} = TAR + SAR$ *Ø=0 scatter component*

T200.   A patient is treated on a cobalt-60 unit with an anterior neck field and a midline cord block according to the diagram shown below. The dose to the cord (at a depth of 8 cm) due to scatter radiation only will be:

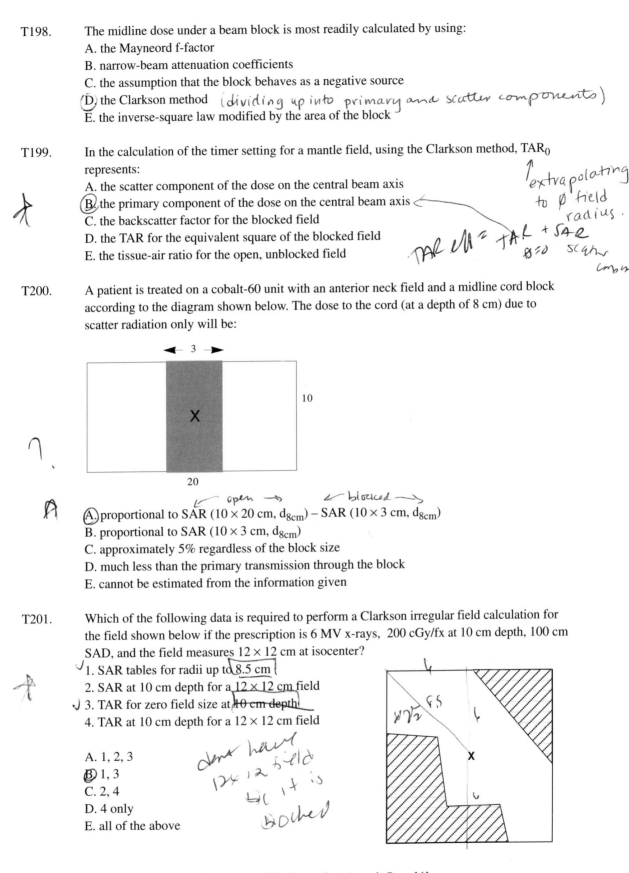

A. proportional to SAR (10×20 cm, d_{8cm}) – SAR (10×3 cm, d_{8cm})   *open / blocked*
B. proportional to SAR (10×3 cm, d_{8cm})
C. approximately 5% regardless of the block size
D. much less than the primary transmission through the block
E. cannot be estimated from the information given

T201.   Which of the following data is required to perform a Clarkson irregular field calculation for the field shown below if the prescription is 6 MV x-rays, 200 cGy/fx at 10 cm depth, 100 cm SAD, and the field measures 12×12 cm at isocenter?

1. SAR tables for radii up to 8.5 cm
2. SAR at 10 cm depth for a 12×12 cm field
3. TAR for zero field size at 10 cm depth
4. TAR at 10 cm depth for a 12×12 cm field

A. 1, 2, 3
B. 1, 3
C. 2, 4
D. 4 only
E. all of the above

dont have 12×12 field bc it is blocked

T202. In an irregular field calculation, the increase in timer (or MU) setting to account for blocking will be greatest for:

A. 10 MV photons, at a depth of d_{max}
B. cobalt-60 photons, at a depth of d_{max}
C. 10 MV photons, at a depth of 10 cm
D. cobalt-60 photons, at a depth of 10 cm
E. 4 MV photons, at a depth of 5 cm

lowest energy greatest depth
more scatter lower energy

T203. Factors that affect the midline dose under a beam block, such as used in a mantle field, are:

1. Block thickness
2. Area of block
3. Unblocked area of field
4. The spectrum of incident x-rays

A. 1, 3
B. 2, 4
C. 1 only
D. 1, 2, 3
E. 1, 2, 3, 4

T204. If the HVL of cobalt-60 is 1.2 cm of lead, a 6 cm block will transmit _____ % of the primary beam.

A. 1.6
B. 3.0
C. 5.0
D. 6.25
E. greater than 6.25

$\frac{1}{2^5} \approx 3\%$

$\left(\frac{1}{2}\right)^5 = \frac{1}{32} \times 100\%$ $\frac{6}{1.2} = 5$

T205. The HVL in lead for cobalt-60 is 1.2 cm. The dose under a cord block, 1 cm wide and 5 cm thick, on a cobalt-60 unit, measured at 5 cm depth in a phantom is about ___% of the dose without the block.

A. 1.6
B. 4
C. 20
D. 0.6
E. 40

$\frac{1}{2^{5/1.2}} \approx 6\%$ + extra scatter from sides

T206. If a Cerrobend block 7.5 cm thick yields 3% transmission in a 6 MV x-ray beam, then a 6 cm thick block will transmit approximately:

A. 1.5%
B. 4%
C. 5%
D. 6%
E. 7.5%

$.03 = \frac{1}{2^x}$ $x \approx 5$

2 HVL = 1.5 cm

$\frac{1}{2^4} \approx 6\%$

PD24. "Similar Triangle" Geometry Problems

T207. A field of 75×50 cm (on the skin) is required to treat a hemi-body patient. The maximum collimator setting is 40×40 cm. The patient must be placed at a minimum of _____ cm SSD. (Standard SSD on this unit is 100 cm; assume no collimator rotation.)

A. 164.5
B. 125.0
C. 210.0
D. 187.5 ✓
E. 155.6

(handwritten:) $\frac{75}{40} \times 100\, SSD_1 = SSD_2$

$75 \cdot 100 = 40 \cdot x$

T208. A simulator film is taken with the geometry shown below. The field measures 15×15 cm on the film. A block 3 cm wide is also drawn on the film. On a tray at 67 cm from the source, the block width would be:

A. 1.5 cm
B. 1.9 cm
C. 2.3 cm
D. 3.0 cm
E. 0.5 cm

(handwritten:) $\frac{3}{130} = \frac{x}{67}$ $x = 1.55$

$\frac{3}{130} = \frac{x}{67}$

100 cm

20 cm

10 cm

Table

Film

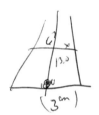

T209. A simulator film is taken with the geometry shown in the question above. The field measures 15×15 cm on the film. The field size on the patient's surface is _____ cm:

A. 13.6×13.6
B. 11.5×11.5
C. 12.5×12.5
D. 8.9×8.9
E. 16.5×16.5

(handwritten:) $\frac{15}{130} = \frac{x}{100}$ $x = 11.5$

T210. A 10×10 cm field is treated at 80 cm SSD. The field size at a depth of 7 cm is ___ cm.

A. 9.2×9.2
B. 10×10
C. 10.9×10.9
D. 11×11
E. 12.1×12.1

(handwritten:) $\left(\frac{87}{80}\right)(10) = 10.9$

$\frac{10}{80} = \frac{x}{87}$

T211. Parallel opposed fields are set up on a cobalt-60 unit at 80 cm SAD. The patient's AP thickness is 24 cm, and the field size at midplane is 15×15 cm. The field size on the skin is ___ cm.
A. 10.8
B. 12.8
C. 13.8
D. 15.0
E. 17.6

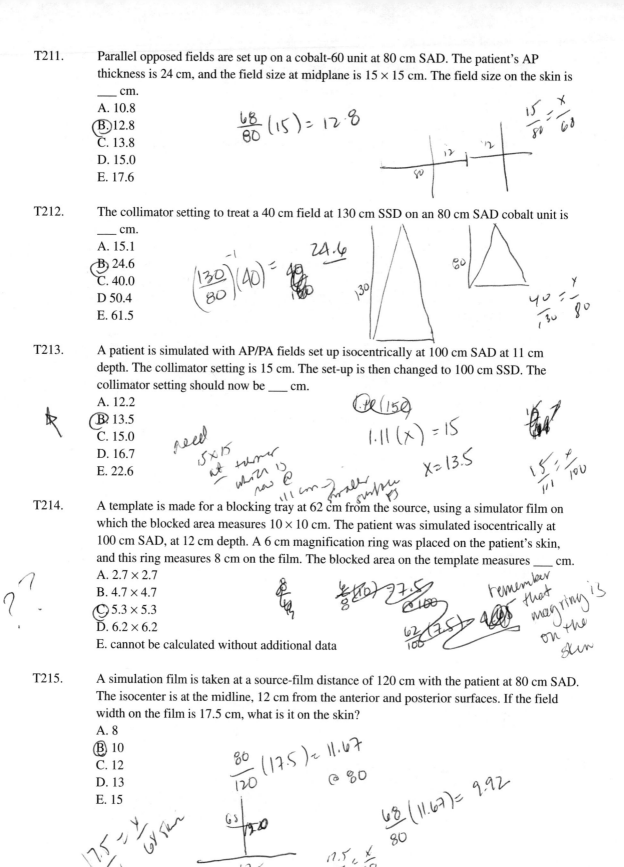

$$\frac{68}{80}(15) = 12.8$$

$$\frac{15}{80} = \frac{x}{68}$$

T212. The collimator setting to treat a 40 cm field at 130 cm SSD on an 80 cm SAD cobalt unit is ___ cm.
A. 15.1
B. 24.6
C. 40.0
D 50.4
E. 61.5

$$\left(\frac{130}{80}\right)^{-1}(40) = \quad 24.6$$

$$\frac{40}{130} = \frac{Y}{80}$$

T213. A patient is simulated with AP/PA fields set up isocentrically at 100 cm SAD at 11 cm depth. The collimator setting is 15 cm. The set-up is then changed to 100 cm SSD. The collimator setting should now be ___ cm.
A. 12.2
B. 13.5
C. 15.0
D. 16.7
E. 22.6

need 5×15 at tumor which is @ 11 cm

$1.11(x) = 15$

$x = 13.5$

$$\frac{15}{11} = \frac{x}{100}$$

T214. A template is made for a blocking tray at 62 cm from the source, using a simulator film on which the blocked area measures 10×10 cm. The patient was simulated isocentrically at 100 cm SAD, at 12 cm depth. A 6 cm magnification ring was placed on the patient's skin, and this ring measures 8 cm on the film. The blocked area on the template measures ___ cm.
A. 2.7×2.7
B. 4.7×4.7
C. 5.3×5.3
D. 6.2×6.2
E. cannot be calculated without additional data

$\frac{6}{8}(10) = 7.5$ @ 100

$62(7.5) = 4.65$

$\frac{62}{100}$

remember that mag ring is on the skin

T215. A simulation film is taken at a source-film distance of 120 cm with the patient at 80 cm SAD. The isocenter is at the midline, 12 cm from the anterior and posterior surfaces. If the field width on the film is 17.5 cm, what is it on the skin?
A. 8
B. 10
C. 12
D. 13
E. 15

$$\frac{80}{120}(17.5) = 11.67 \quad @ 80$$

$$\frac{68}{80}(11.67) = 9.92$$

$\frac{17.5}{120} = \frac{Y}{68}$ skin

$\frac{17.5}{120} = \frac{x}{68}$

T216.   A medulloblastoma patient is set up on a linac at 100 cm SSD with a field size of 6×42 cm. The patient is then transferred to a cobalt-60 unit which has a maximum field size of 35 cm at 80 SAD. What is the minimum SSD required?
A. 80 cm
B. 104 cm
C. 96 cm
D. 115 cm
E. 100 cm

(handwritten: $\frac{42}{100} = \frac{35 \text{ max field}}{80}$ size; $\frac{42}{35} = 1.2$; $1.2(80) = 96$)

T217.   A patient was simulated for treatment on a linear accelerator at 100 cm SAD for a 10×10 cm field but has been scheduled for treatment on a cobalt-60 at 80 cm SAD. If blocks need to be cut for the cobalt-60 unit from the 100 cm film that has a target-to-film (TFD) distance of 150 cm, what TFD should be used? (Assume that the field size does not need to be adjusted.)
A. 120 cm
B. 130 cm
C. 150 cm
D. 188 cm
E. cannot use same simulator field, must re-simulate

(handwritten: $\frac{.50}{.70} = \frac{x}{80}$; $\frac{80}{100}(150) = 120$)

T218.   All of the following can be used to determine the magnification of a simulator film *except*:
A. a grid of lead BB shot in a special tray that is imaged on the film
B. the target to film distance when the SAD is 100 cm
C. field defining wires that are imaged on the film, when the field size is known
D. magnification ring placed on the patient
E. a steel ruler placed on the patient

PD25. SSD Versus SAD Set Up

T219.   A patient's mediastinum is being treated isocentrically by an AP/PA technique with 16 MV x-rays. The linac requires service, so the patient is transferred to a cobalt-60 unit. All of the following are true *except*:
A. the field size at midplane will change because the cobalt-60 has an SAD of 80 cm
B. the cord dose will increase
C. the skin dose will increase
D. the field size on the skin will decrease *b/c of increased divergence*
E. the dose variation throughout the treatment volume will increase.

(handwritten: but projection will be smaller; divergence @ 80 SAD)

T220.   A patient is usually treated at 100 cm SSD. If the patient must be treated on a stretcher at 135 cm SSD, which of the following statements is *not* true?
A. the collimator setting will decrease
B. the monitor unit setting will increase
C. the exit dose will increase *b/c of increased PDD*
D. the greatest factor affecting the change in the MU setting is the Mayneord's f-factor correction to the PDD

(handwritten: greatest factor will be the $\frac{1}{4^2}$ factor not Mayneord's factor)

T221.   A patient is simulated at 80 cm SAD for treatment on a cobalt-60 machine. It is then decided to treat the patient on a 100 cm SAD linac. Which of the following will be different?
A. field size at the isocenter as shown by the field edge wires on the simulator film
B. triangulation points
C. depth to isocenter
D. collimator settings *will be the same b/c both @ SAD*
E. skin marks showing field edges

PD26. Miscellaneous

T222.   Integral dose-volume histograms for the lung for two treatment plans, A and B, are shown below. Which of the following statements is *false*?

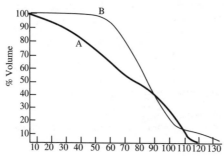

A. approximately the same lung volume receives at least 80% of the prescribed dose for both plans
B. plan B gives a small lung volume (<10%), a dose significantly greater than the prescribed dose
C. dose-volume histograms give information about the position of high dose regions within the organ
D. dose-volume histograms can be generated on some treatment planning computers for target volumes or critical structures

T223.   Basic dosimetry measurements to be done prior to implementing total body irradiation (TBI) techniques include:
A. central axis dose calibration made under conditions representative of the actual treatment geometry
B. central axis data such as percent depth dose or tissue-maximum ratios measured with treatment geometry
C. test of the inverse square law over the range of treatment distances
D. dose profiles of the flatness and symmetry of the beam
E. all of the above ·

T224.   Features of a true 3-D treatment planning system that are generally *not available* in a 2-D system are all of the following *except*:
A. dose-volume histograms
B. isodose surface display
C. calculations that account for scatter from inferior or superior CT slices
D. inhomogeneity corrections
E. calculation over a volumetric grid of dose points

T225.   All of the following are used for stereotactic radiosurgery *except*:

A. linear accelerators
B. gamma knife cobalt-60 unit
C. protons
D. electrons

T226.   Factors affecting the dose rate at d_{max} on a cobalt-60 unit are:

1. field size
2. SSD
3. patient thickness
4. source activity
5. given dose
6. timer error

A. 1, 2, 3, 4, 5, 6
B. 1, 2, 4
C. 2, 3, 4
D. 1, 2, 4, 5
E. 2, 4, 5, 6

T227.   A circle of diameter 7 cm is drawn on a prostate contour to represent the treatment volume. For a plan with three oblique fields set up isocentrically, the field width chosen for the treatment fields would be:

A. 6.0 cm
B. 6.5 cm
C. 7.0 cm
D. 7.5 cm
E. 8.0 cm

Answers

PD1.

T1.   E   Scatter is associated with Compton interactions. At 2 mm Al HVL there is a considerable amount of photoelectric interaction. Above about 1 mm Cu HVL, photoelectric interactions are negligible and Compton interactions predominate. As the photon energy increases more energy is transferred to the electron and less to the scattered photon. The maximum backscatter is at about 0.7 mm Cu HVL.

T2.   D   TMR, like TAR, is independent of SAD.

T3.   D   The timer or MU setting for rotation uses the average TAR, averaged over all depths for the area of rotation.

T4.   A   As SSD increases PDD, and thus exit dose, increase (per Mayneord's f-factor).

T5.   C   (A) TAR is independent of SSD or SAD. (B) BSF increases with increasing energy up to about 1.0 mm Cu HVL, then decreases as energy increases. (D) Although historically TARs were measured for cobalt-60 and TMRs were introduced for higher energy beams, TMRs can be measured and used for calculating timer settings at any megavoltage energy.

T6.   D

T7.   E

T8.   B

T9.   E   TMR = tissue-maximum ratio. It is the ratio of two dose rates measured in phantom, at the same distance from the source; one with a chosen thickness of overlaying phantom, the other with only the thickness required to attain d_{max}. It is independent of SSD, and can be measured on any megavoltage x-ray or gamma ray unit.

T10.   A   The tissue-air ratio is the dose rate at depth divided by the dose rate in air at the same point. The TAR at d_{max} is called the back scatter factor (BSF). BSF decreases as energy increases above 1 MV.

T11.   C   PDD increases with increasing SSD, since it contains an inverse square component as well as attenuation. TMR, TAR and BSF (TAR at d_{max}) measure attenuation only, and are independent of SSD.

T12.	E	
T13.	D	
T14.	A	
T15.	B	These answers are essentially the definitions of the quantities indicated. Note that the "dose in air" refers to the dose to a small mass of material of radius d_{max}.
T16.	B	The dose to point A will be about 4% lower than that found from tables, due to the lack of total backscatter. This fact is generally ignored in treatment planning dose computation algorithms.
T17.	E	The backscatter factor is a function of beam quality and field size. It increases to a value of about 1.5 for a large field at about 1 mm Cu HVL and then falls to a negligible value above about 10 MV. It cannot have a value of 1.0 (answer A) because this would imply that there is no difference between the dose rate in air and that at d_{max} in tissue.
T18.	C	
T19.	B	
T20.	D	Tissue-maximum ratio (TMR) is independent of SSD since the dose at depth and the dose at d_{max} are measured at the same distance from the source.
T21.	D	
T22.	B	As the beam energy increases, the depth of d_{max} increases and the size of the chamber build-up cap for the in-air measurement will also increase. A large size build-up cap will start to act as a "mini" phantom.
T23.	A	Scatter from the corners of the rectangle has farther to travel than from the periphery of the square and will contribute less to the depth dose.
T24.	D	TAR is independent of SSD.
T25.	C	TMR at d_{max} is 1.0 by definition for any photon energy.
T26.	A	
T27.	A	TMR = (Dose rate at depth) / (Dose rate at d_{max}) = (Dose rate at depth) / (Dose rate in air × BSF) = TAR / BSF
T28.	E	PDD increases with increasing SSD, according to Mayneord's f-factor, because the inverse square factor becomes relatively less important at greater SSD.

PD2.

T29.	D	The beam-on time can be found by dividing the prescribed dose of 150 cGy by the dose rate times the PDD. t = 150/(100 × 0.65) = 2.3 min.

T30.   A   A field size of 10 cm at the tumor (8 cm depth) requires a field size of 10 × (80)/(88) = 9 cm on the surface at 80 cm SSD. The equation used to calculate the timer setting for an SSD set up is: time (t) = ((D/(DR × PDD)) + timer error where D = dose at prescription point; DR = Dose rate at d_{max}. At treatment SSD DR = DR air at 80.5 cm × BSF for field size at the skin. PDD = PDD for field size at the surface and depth to prescription point.

T31.   D   Timer setting = dose at depth/(dose rate at d_{max} × PDD) = 300/(100 × 0.78) = 3.85 mins. Given dose = dose at d_{max} = dose at depth/PDD = 300/0.78 = 385 cGy.

T32.   D   The simplest formula to calculate timer setting for a single field at SSD is:
t = dose at depth/[(Dose rate at d_{max}, SSD) × PDD (equ. sq., d)]
Other formulae involving dose rate in air, BSF, inverse square corrections and TAR can be used to arrive at the same answer, but they are less direct. The other formulae in the question are incomplete.

T33.   C   Timer setting = (dose per fraction at d = 4 cm)/(DR × PDD/100) where DR = dose rate (cGy/min) at d_{max}, 80 cm SSD, in tissue. Graphs of dose rate at d_{max} in tissue are usually presented in one of two ways: (1) a graph of dose rate vs. equ. sq. field, which must be decayed by 1.1% and replotted each month, or: (2) a graph of relative output factor vs. equ. sq. field (which is always valid, since it is a set of ratios), and a value of dose rate at d_{max}, 80 SSD for a 10 × 10 cm field, decayed for that month. Thus: DR (9 × 9 cm) = DR (10 × 10 cm) × ROF (9 × 9 cm).

T34.   B   100 × 1.035 × (80/80.5)^2 = 102.2 cGy/min

T35.   D   For 180 cGy/total to the 100% level, the dose per beam to the midplane = 90 × (105/100) = 94.5 cGy. t = 94.5/ (200 × 0.8 × 0.65) = 0.91 mins.

T36.   C   Note that for clinical use, beam data will usually be prepared by the physicist so that it is in the easiest form to use. For example, for 80 cm SSD calculations the dose rate (DR) would be given at d_{max}, 80 cm SSD, in tissue either on a graph of DR vs. equivalent square, or for 10 × 10 cm only to be used with a table of relative output factors (ROFs). The same data can be calculated, however, from an output measured in air, with the appropriate corrections for inverse square, ROF and BSF, as in the calculation below.

Equivalent square = 4 × area/perimeter = 9.2 cm

Time = 300/[(DR$_{air}$ for 10 × 10 cm, 80 SAD) × (80/80.5)^2 × ROF (9.2) × BSF (9.2)]
= 300 / (115.0 × 0.988 × 0.993 × 1.032) = 2.58 mins.

T37.	E	The ROF is for the collimator setting (7.4), but the BSF is for the equivalent square on the skin (9.2 cm). Time = $300/[115 \times (80/100.5)^2 \times 0.980 \times 1.032] = 4.07$ mins.
T38.	B	Time = $100/[\text{DR}_{air}(10 \times 10 \text{ cm}, 80 \text{ SAD}) \times \text{ROF}(12 \times 12 \text{ cm}) \times \text{TAR}(12 \times 12 \text{ cm}, d_{9cm})] = 100 / 115.0 \times 1.013 \times 0.775 = 1.11$ mins.

PD3.

T39.	D	The effect on the TAR of a small corner block can be ignored, but the tray factor (TF) must be included. Time = (dose at isocenter)/[(dose rate in air at 80 cm) $\times$ (TF) $\times$ (TAR)] $= 90/[(120 \times (0.96) \times (0.686)] = 1.14$ mins.
T40.	E	The equation used to calculate the timer setting for an isocentric set up is: time (t) = $[(D/(\text{DR} \times \text{TAR})]$ + timer error where D = Dose at depth of isocenter DR = Dose rate in air for field size at isocenter TAR = Tissue-air ratio for field size at isocenter and depth to isocenter
T41.	E	MU = (dose/fraction)/[Output at SAD $\times$ TMR($d_{12.5cm}$)] = $90/(1.078 \times 0.777) = 107$
T42.	C	t = [(dose at midplane)/(dose rate in air $\times$ TAR $(10 \times 10, d_{11})$)] + timer error $= [(100)/(125.5 \times 0.762)] + 0.02 = 1.07$ mins.
T43.	D	The information required is the same as would be required to calculate a single isocentric field: time = (dose at isocenter)/(dose rate in air $\times$ TAR $\times$ WTF) To calculate the dose per beam at the isocenter: The total dose per fraction at the isocenter = the prescribed dose $\times$ (ratio of isodose through isocenter/isodose at which dose is prescribed).
T44.	D	MU = (tumor dose)/(DR $\times$ TMR) where DR = dose rate at d_{max} in tissue, at 100 cm SAD = dose rate in air, 100 cm SAD $\times$ BSF. MU = $100/1.015 \times 1.044 \times 0.806 = 117$
T45.	C	The relative output factors for a cobalt-60 unit range from about 0.96 (5×5 cm) to 1.07 (30×30 cm). On a linac, the range can be larger due to scatter into the ion chamber from the collimators.
T46.	C	The general time-on formula is: time = (dose at depth/dose rate at depth). Dose rate at depth can be found by either using dose rate in air at SAD $\times$ TAR or dose rate in tissue at SAD at $d_{max} \times$ TMR. (Note: BSF $\times$ TMR = TAR).

very imp →

PD4.

T47.	E	The dose at d_{max} for an SSD calculation can be found by dividing the prescribed dose of 150 cGy by the PDD. D = $150/0.65 = 231$ cGy.

T48.	A	Exit dose to cord = given dose to s'clav × PDD. PDD is a function of field size and cord depth.

T49.	A	300 cGy × 70%/90% = 233 cGy.

T50.	C	Dose at d_{max} = (D_{mid}/2) × (1.0/0.7) = (D_{mid}/2) × 1.43 Exit dose at this point = (D_{mid}/2) × (PDD_{18}/PDD_{10}) = (D_{mid}/2) × (0.45/0.7) = (D_{mid}/2) × 0.64 Total dose at d_{max} = (D_{mid}/2) × (1.43 + 0.64) = 1.035 D_{mid}.

T51.	C	Dose (B) = Dose (A) × [PDD (10 × 10 cm, d_{6cm}) / PDD (10 × 10 cm, d_{9cm})] × (80.5/83.5)^2.

PD5.

T52.	D	The total dose at d_{max} cannot be equal to or less than the midplane dose. D and E are 10% and 25% greater than the midplane dose; for 8 MV x-rays D is the only reasonable answer. The general formula for finding dose at d_2 from dose at d_1, for an isocentric set up, is: $D(d_2) = D(d_1) \times (TMR_2/TMR_1) \times (SAD_1/SAD_2)^2$ Entrance dose at d_2 = 90 × (1.00/0.777) × (100/89.5)^2 = 144.6 cGy Exit dose at d_{23} = 90 × (0.564/0.777) × (100/110.5)^2 = 53.5 cGy Total dose = 144.6 + 53.5 = 198 cGy, which is 10% more than the midplane dose. Note: the most common mistake is to omit the inverse square correction; the TMR ratio accounts for the difference in attenuation only, but as the points are at different distances from the source, an inverse square correction is also required.

T53.	B	Two factors are required to calculate the dose at Q relative to P: (1) an inverse square factor, and (2) a tissue attenuation factor. Dose at Q = Dose at P × (100/102)^2 × 1.04 = Dose at P × 1.0 In this case the inverse square and attenuation factors cancel each other.

T54.	D	$D_{neck} = D_{CAX} \times \dfrac{TMR\ (d_5)}{TMR\ (d_8)} = 4000 \times \left(\dfrac{.945}{.879}\right) = 4300$ Although the midpoint at the neck is 98 cm from the anterior and 102 cm from the posterior, the inverse square corrections for the anterior and posterior doses cancel out.

PD6.

T55.	C	Note: the percentage at d_{max} will be greater for smaller fields, and less for larger fields due to the change of PDD with field size.

T56.	D	The increase in cord dose over midplane dose on the axis will be greatest for lower photon energy, and for the greatest difference between AP thickness at the neck and at the beam axis. Also, for AP/PA fields, the closer the point is to d_{max}, the greater the increase over the midline dose.

T57.   C   The maximum tissue dose is at d_{max}, and is the given dose (GD) + exit dose at depth 19.5 cm. GD = (dose at midplane/2 × PDD at d_{10cm}) = 1500/0.525 = 2857 cGy
Exit dose = dose at midplane × [(PDD at $d_{19.5cm}$)/(PDD at d_{10cm})] = 1500 × (0.241/0.525) = 689 cGy
Total dose at d_{max} = 3546 cGy.

T58.   D   The lowest dose to points between midplane and the surface results from using the highest energy and largest SSD, to maximize PDD.

T59.   E   Any factor that increases PDD will decrease the total dose at d_{max}, compared with the total dose at midplane. Treating at SSD rather than SAD gives slightly higher PDDs.

T60.   C   It is useful to plot graphs of dose to d_{max} (as a % of midplane dose) for several field sizes vs. patient separation, for all photon energies used in a department. This facilitates deciding which patients should be treated with which energy, and when a four field technique should be used.

T61.   A   The rule of thumb for cobalt-60 is 10% extra dose at d_{max} for 20 cm separation (for SSD set-up), although this will vary somewhat with field size. Note that for d_9, the value is about 8% for the SSD set-up, as opposed to 11% for the SAD set-up. This reflects the increase in PDD as SSD increases.

T62.   C

T63.   D

T64.   D   Total dose at d_{max} is slightly higher for SAD set-up, because of the shorter SSD. Isocentric treatment has less set-up and treatment time, and all fields have the same height at the isocenter. (Fields set up at SSD should have different heights at SSD, to project to the same height at midplane).

PD7.

T65.   B   Using Mayneord's f-factor:
PDD(80 SSD, d_{10cm}) = PDD(100 SSD, d_{10cm}) × $(110/100.5)^2$ × $(80.5/90)^2$ = 71.9%

T66.   C   PDD at SSD_2 = (PDD at SSD_1) × F^2
Where F = {[(SSD_1 +d)/(SSD_1 + d_{max})] × [(SSD_2 + d_{max}) / (SSD_2 + d)]}
F is independent of photon energy. For SSD_2 > SSD_1, F > 1.0
f is the roentgen to cGy conversion factor, 0.957 for cobalt-60.

T67.   E   Mayneord's f-factor is based on a strict application of the inverse square law, without regard to changes in scattering as the SSD is changed.

$$F = [(f_2 + d_{max})/(f_1 + d_{max})]^2 × [(f_1 + d)/(f_2 + d)]^2$$
f_1 = original SSD, f_2 = new SSD
d = treatment depth

T68. B The PDD at 200 cm SSD is found by multiplying the PDD at 100 cm SSD (for the same field size on the skin) by Mayneord's f-factor. This factor simply removes the inverse square component of the PDD for 100 cm SSD, and inserts the inverse square factor appropriate for the new SSD of 200 cm.

PDD $(30 \times 30$ cm, 200 SSD$)$ = PDD $(30 \times 30$ cm, 100 SSD$) \times F^2$

where F = {[(std SSD + d) × (new SSD + d_{max})]/

[(std SSD + d_{max}) × (new SSD + d)]}

PDD (200 SSD) = $0.730 \times (110/101.5)^2 \times (201.5/210)^2 = .789$

PD8.

T69. D The equivalent square of a rectangular field is side 4 × area/perimeter. The area (A) of the field is $9 \times 17 = 153$ cm^2. The perimeter (P) is $2 \times (9+17) = 52$ cm.

$(4 \times A)/P = 4 \times 9 \times 17/2 \times 26 = 11.76$.

T70. B The equivalent square of a rectangular field is that square field which has the same scatter contribution on the axis, and hence the same PDD and TAR. One square centimeter near the axis contributes more scatter than the same area at the corner of the field. Hence the equivalent square of a rectangle has a smaller area than the rectangle.

T71. B An approximate rule for the side of the equ. sq. is $(4 \times A)/P$ where A = area, P = perimeter.

T72. A

T73. A BSF = TAR at d_{max}.

T74. B The equivalent square field is approximately C × C cm, where C = 4 × area/perimeter.

$4 \times A/P = (4 \times 15 \times 22)/ 2(15 + 22) = 17.8$ cm.

T75. A Scatter from the corners of the rectangle has farther to travel than from the periphery of the circle and will contribute less to the depth dose.

PD9.

T76. B Positioning of the wedges is not limited to 90°.

T77. B Uniform dose distributions can be obtained in many cases with only two wedged fields. A third unwedged field is sometimes used to good effect with a pair of wedges.

T78. A This is the ideal relation between HA and WA if compensation for curvature is not required.

T79. A For almost all treatments using wedges, the "heels" should be together.

T80.   D   Wedge angle (WA) = 90°– (Hinge angle/2). WA = 90°– (60/2) = 60°.
The appropriate wedge will provide isodose curves which run parallel to the bisector of the hinge angle.

T81.   E   The wedge angle is the angle through which the 50% isodose curve is turned (from its position in the open field).

T82.   A   Increasing the wedge angle would attenuate the anterior portions of the lateral fields, which would decrease the dose gradient. B and C would have the opposite effect.

T83.   E   Universal wedges are used by some linac manufacturers to allow any wedge angle to be achieved by a suitable combination of weights for the wedged and open fields. Equal doses of open and 30° wedged beams will create the effect of a 15° wedge; however, the time or MU for the wedged field will be greater by l/WTF.

T84.   B   Timer setting = (dose at depth) / [(dose rate at d_{max}) × (WTF) × (PDD)] = 150/ [(100) × (0.75) × (0.68)] = 2.94 min.

T85.   A   Dose per fraction = 180 cGy; dose per field = 60 cGy. MU = (dose per beam) / (output × WTF × TAR) = 60 / (0.9 × 0.58 × 0.782) = 147 MU

T86.   C   MU = (dose at depth)/ (dose rate at depth)
where dose rate at depth = dose rate at d_{max} (open) × PDD × WTF. Thus MU = 150/0.59 = 254.

T87.   D   The dose on the beam axis will be the same, but the dose distribution off axis will be different. A 60° wedge could be used (provided its WTF is used in the timer setting calculation) to restore the total dose, after a total of 10 treatments, to the planned distribution. However, the radiobiological effect will not be the same, because of the different doses per fraction to different areas.

PD10.

T88.   C   Perpendicular beams (hinge angle 90°) are best matched with 45° wedges when the beams enter perpendicularly to the surface. Usually, however, the surface is curved, and the wedge must also compensate for the "missing tissue."

T89.   D   In a "wedged pair," the thick ends of the wedges are always together.

T90.   B   The isodose distribution is underwedged and a larger angle wedge should be used. If the distribution is renormalized it is not improved, rather, the gradient across the breast will still be 20% no matter where the 100% level appears.

T91.   B   Wedges act as compensators for sloping surfaces and increase homogeneity for the target volumes. If the wedge angles are properly chosen, homogeneity is increased when using an orthogonal pair. Bilateral opposing wedged fields create a gradient anterior to posterior and a properly weighted anterior open field will combine with the gradient for a more homogeneous plan. The wedges act as compensators for a larynx field and will reduce the anterior hot spot.

T92.　D　Wedged beam data can be stored normalized to 100% at d_{max} on the central axis, or 100% × WTF. This depends on the convention of the treatment planning system, and sometimes on the preference of the physicist entering the data. Thus, two plans with "equally weighted beams" can be very different, depending on the above convention, and also on the precise definition of "weight," which can also differ between treatment planning systems.

PD11.

T93.　D　A "hand" calculation to check the MU setting would use the following: MU = (total dose at isocenter)/(output in air × TAR) or (total dose at isocenter)/(output at d_{max} × TMR) where the TAR is for the field size at the isocenter, and the average depth to the isocenter. The average of the AP, PA and lateral depths gives a reasonable estimate of average depth.

T94.　B　Time = $100/[DR_{air} (10 \times 10 \text{ cm}, 80 \text{ SAD}) \times ROF \times$ TAR $(12 \times 12 \text{ cm}, d_{17.5})] = 100/[115.0 \times 1.013 \times 0.500] = 1.72$ min.

T95.　B　A 180° arc gives an isodose distribution in which the dose falls off across the volume. If the field size used is twice the width of the volume and the isocenter is "past-pointed," the dose homogeneity is acceptable, but the integral dose is unnecessarily high.

T96.　D　A good approximation of TAR is the TAR for the sum of the AP and lateral separations divided by four. $(22+32)/4 = 13.5$ cm, TAR (13.5 cm) ~ 0.60.

PD12.

T97.　C　Surface dose decreases as photon energy increases. The reverse is true for electrons.

T98.　B　As SSD decreases scatter from the collimators will increase. Skin dose increases as field size increases. Bolus is generally used to remove skin sparing and bring the skin dose up to 100%. Oblique incidence causes secondary electrons to travel in a more parallel, rather than perpendicular, direction to the skin. The buildup depth is reduced, and skin dose increases.

T99.　E　The range of dose rates obtainable on treatment units (70 cGy/min for a low activity cobalt-60 source up to 500 cGy/min on some linacs) is not known to have any effect. A-D, however, are reasons for a higher skin dose in the neck.

T100.　E　A-D will all increase skin dose.

PD13.

T101. **C** Even if fields with unequal field sizes are properly gapped, the larger AP field will diverge into the smaller PA field (as will the larger PA into the smaller AP). This will create hot and cold spots on either side of the junction.

T102. **D** Calculating a gap on the skin so that the 50% isodose lines of both fields match at treatment depth is the most common way to eliminate hot spots at the junction. Angling the gantry so that both fields abut will eliminate divergence at the beam edges and the 50% lines can be matched. A penumbra generator is a wedge that can be custom designed to broaden the penumbra so that 50% lines can be matched and positioning is less critical. A collimator angle won't help for adjacent fields but can help when matching orthogonal fields.

T103. **D** The gap for each field can be found from the formula: $g = (d/SAD \times FS/2)$ where d = depth of matching point, in the region of the junction, and FS = field size at SAD. $G = g_1 + g_2 = (6/80) \times (20/2) + (6/100) \times (30/2) = 1.65$ cm.

T104. **D** For the initial field configuration at 80 cm SSD, the gap will be 2 cm. To achieve field abutment with 100 cm SSD, the gap would be 1.6 cm, a reduction of 0.4 cm.
$Gap = depth \times (L_1 + L_2)/2 \times SSD$

T105. **C** $Gap = d \times (C_1 + C_2)/ (2 \times SAD) = 5 \times (20 + 22)/ (2 \times 80) = 1.3$ cm.

T106. **B** $Gap = d/SAD (C_1/2 + C_2/2) = 9/100 (20/2 + 24/2) = 1.98$ cm.

T107. **A** The geometry of the gap formula relies on all beam edges meeting at a point, where each contributes 50% of the dose on the axis. Thus the total dose at the "match" point is equal to that on the axis of either pair of fields. Since the edge of the light field is used to represent the edge of the radiation field, it is important that the light field matches the point representing 50% of the dose on the axis at this level. Provided the geometry is correct, different energy beams, or different treatment machines may be used for each field. For parallel opposed pairs of fields, the *best* match is obtained when the beams diverge at equal angles, i.e., when the collimator settings are equal.

T108. **A** The general formula for calculating a gap is:
(Field size/2) $\times$ (depth at the junction/SAD)
Field I: $(20/2) \times (12/100) = 1.2$ cm
Field II: $(30/2) \times (12/80) = 2.25$ cm
Total gap = 1.2 + 2.25 = 3.45 cm

109. **D** To match the divergence of the larger field and the smaller field, the SSD of the second field can be adjusted so that the collimator settings are the same, but the field size projected at the skin is 25 cm as required. By similar triangle geometry:

$$\frac{\text{field size 1}}{\text{SSD}_1} = \frac{\text{field size 2}}{\text{SSD}_2}$$

$$\Rightarrow \text{SSD}_2 = \left(\frac{\text{field size 2}}{\text{field size 1}}\right) \times \text{SSD}_1$$

$$= \left(\frac{25}{20}\right) \times 100 = 125 \text{ cm}$$

T110.　D　　A, B and C will remove divergence at the chest wall border; collimator rotation will not do this, although it may be required in order to align the fields with the superior/information slope of the chest wall.

T111.　B　　Collimator angle = c degrees, where tan c = 0.5 × field height /SSD = 12.5 / 100
c = $\tan^{-1}$ (0.125) = 7°

T112.　B　　The head fields are rotated, superior end up, to match the divergence of the upper spine field. Divergence = $\tan^{-1}$ [(0.5 × collimator height)/SAD] = 10°.

PD14.

T113.　A　　d_{max} for cobalt-60 is about 0.5 cm for a 10 × 10 cm field but decreases rapidly as the field increases.

T114.　E　　d_{max} is negligible for all superficial x-rays.

T115.　D　　d_{max} increases as the energy of the beam increases.

T116.　D　　The depth at which electron equilibrium is achieved (d_{max}) increases with increasing energy for both photons and electrons.

T117.　B

T118.　D

T119.　A

T120.　E

T121.　A

T122.　A

T123.　C

T124.   A

T125.   B

T126.   E

T127.   E

T128.   D   The highest energy x-rays will always be the most sparing of normal tissue, when treating the pelvis.

T129.   D   Lung corrections are less for high energy photon beams, where the attenuation per cm of tissue is less. The opposite is true for A and B, and in C, the flatness of most linacs is defined at 10 cm depth, so there should be no difference between the two beams.

T130.   C   A rough rule of thumb for cobalt-60 is that the percent depth dose decreases 4% per cm (PDD at 12 cm is 50%). The PDD at 5 cm is 80%, and PDD at 6 cm 76%. The PDD value is the only significant difference when calculating the time-on for this situation, so the magnitude of error is 80/76 = approx. 5% error.

T131.   D   If the depth dose on a cobalt-60 unit changes approximately 4% per cm, the PDD at 10 cm is 60% and the PDD at 8 cm is 68%. The time-on would be calculated with a depth dose of 60% originally and should be calculated with 68% when the separation changes, making the original time-on too long by 68/60 = 1.13; greater than 10%.

PD15.

T132.   E   Geometrical penumbra = source diam. × (SSD-SCD) /SCD where SCD = source – collimator distance. As photon energy increases, the energy of scattered electrons increases, so penumbra is generally wider for higher energy beams.

T133.   A   High energy x-ray beams interact primarily by the Compton and pair production processes. Both processes result in the production of scattered photons having energies lower than the energies of the incident photons and traveling in considerably different directions. These scattered and annihilation photons may be absorbed outside of the geometrical confines of the irradiated volume.

T 134.   A

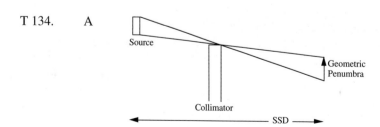

T 135. D The geometric penumbra (w) of any ionizing photon beam is equal to: w = c [(SSD-SCD)/(SCD)] where c is the diameter of the source, SSD is the source skin distance and SCD is the source collimator distance. The target size of a cobalt-60 teletherapy unit is 1.0 to 2 cm. A linac target is typically 3 mm. The difference in target size overwhelms the difference in SSD and SCD. The penumbra is independent of the output and of the effects of the flattening filter. The flattening filter is used to make the intensity distribution relatively uniform across the field.

T136. D A, B and C would decrease the penumbra; E would have no effect.

T137. C Penumbra decreases as source-block distance increases.

T138. C The "horns" of the beam refer to the increase in dose towards the edges of a large field at d_{max}. The flattening filter is thicker on the beam axis than at the edges. This difference in filtration means a variation in the energy spectrum across the beam. The filter cannot achieve a flat beam at all depths, and is generally designed to give a flat beam at 10 cm. This results in horns at d_{max}, and under-flattening at greater depths. Note that the size of the "horns" is a more sensitive monitor of beam energy constancy than PDDs on the axis, and should be checked when energy changes are suspected, e.g., after changing a magnetron.

T139. B Both geometrical penumbra and patient scatter increase with depth, and both contribute to broadening the penumbra.

PD16.

T140. A Integral dose is a measure of the energy deposited. It is calculated by multiplying the dose (in cGys) by the mass of tissue receiving that dose (grams). Since the usual units will be kilocGys × kilograms, the unit of integral dose is the megagram cGy. (Mgm – cGy). This is a unit of energy.

T141. A

T142. B Integral dose should be minimized to reduce side effects of treatment.

T143. B As beam penetrability increases, less dose can be given to surrounding normal tissue for the same tumor dose.

T144. C Integral dose is generally minimized by using the shortest depth from surface to isocenter. For deep tumors a higher photon energy will reduce the integral dose. A low integral dose means a lower overall dose to normal tissues; however, other considerations such as dose to sensitive organs and feasibility of set-up must also be considered when selecting the best treatment plan.

T145. D Lower energy reduces the exit dose, and hence the integral dose.

PD17.

T146.   B   If the patient is 2 cm closer to the source than intended, he or she will receive a greater dose. By the inverse square law the dose will be $(100/98)^2 = 1.04$ of the dose intended, i.e., 4% more.

T147.   B   The inverse square law applies to dose rates in air for the same collimator setting. Dose rate at 200 cm = DR at 80 cm $\times (80/200)^2$

T148.   D   By the inverse square law:
Dose rate$_1$ / Dose rate$_2$ = (dist$_2$)^2/(dist$_1$)^2
therefore, $250/100 = $ (dist$_2$)^2 / $(100)^2$
and dist$_2$ = $(25{,}000)^{1/2} = 158$ cm

T149.   E $207 \times (140/120)^2 = 282$ MU
The field size factor will not change very much, and the effective TMR will be the same. An inverse square correction is required to adjust the output from 120 to 140 cm SAD.

T150.   D   By the inverse square law, the dose at d_{max} will be reduced by $(20/22)^2$; or 17%. Inverse square corrections can be quite large for short SSD units.

PD18.

T151.   A   Adding bolus to the patient to correct the contours eliminates the skin sparing effect of high energy radiation. Tissue compensation filters correct patient contours without loss of skin sparing.

T152.   A   As any object placed in the path of the beam is moved further from the patient, its effect on the surface dose decreases since the electrons produced in the object have a greater chance of scattering out of the beam. Selection B is incorrect since increasing the spoiler thickness would increase the surface dose. Selection C would be correct only if the spoiler was in contact with the patient's skin. Beam spoilers are not generally necessary for treating Hodgkin's disease with energies below 10 MV since the buildup characteristics of the lower energy beams adequately treat the superficial nodes.

T153.   C   Tissue compensators are used to enable a more uniform dose distribution to be delivered. They are placed away from the patient's surface (at the blocking tray distance for instance) and will therefore preserve skin sparing.

T154.   A

T155.   B

T156.   A

T157.   E   Skin sparing is reduced by bolus, a plaster cast, a beam spoiler, and many times a skin reaction is seen in a skin fold since the tissue being treated on top of the fold acts as bolus.

T158.   C   This is sometimes required if only a high energy photon beam is available, which would underdose the superficial aspect of the treatment volume. It decreases the depth of d_{max}, while still maintaining some skin sparing (unlike bolus).

PD19.

T159.   B   The wedge symbol is reversed.

T160.   B   The isodose lines will tend to follow the surface slope.

T161.   B   One wedge is reversed.

T162.   A

T163.   A.

T164.   D

T165.   E

T166.   A

T167.   B

T168.   C

T169.   C   The dose rate around a brachytherapy source falls off very rapidly.

T170.   A

T171.   B

T172.   D

T173.   C   The maximum tissue dose would be $(105/95) \times 200 = 221$ cGy. The hot spots will now be 111%. Renormalization will not change the % variation in dose across the breast; it simply shows the same data in a different way.

T174.   B   Maximum dose is at d_{max} for parallel opposed fields.

T175.   A   Maximum dose is at the isocenter when 4 fields overlap.

T176.　　B　　An arc with open beams of 180° or less will give an uneven dose to the volume around the isocenter. The maximum will be the closest point to the skin which "sees" the beam at all times.

T177.　　C　　In this plan, the small angle between two of the fields would require wedges in these fields (thick ends together) for an even dose distribution around the isocenter. As the beams are open, there will be a hot spot at C.

T178.　　C　　In this breast plan, the most even dose distribution would be achieved with wedged fields. If open fields are used, the maximum dose will be at C, at the point where the air gap is greatest.

PD20.

T179.　　B　　The beam produced by a superficial x-ray machine consists of a spectrum of x-ray energies. Adding filtration to the beam will effectively remove some soft (low energy) photons and improve the quality of the beam, increasing the HVL, and hence the penetrability of the beam.

T180.　　E　　For a prescribed given dose PDD = 100%. The dose rate at the skin surface = the dose rate in air for the cone used × the BSF for the treatment field size.

T181.　　D　　All these factors are required except the PDD; since the dose is prescribed as a "given dose," i.e., on the skin, the PDD is 100%.

T182.　　C　　The HVL (in Al or Cu) defines the penetrability of a low energy x-ray beam. Different combinations of kVp and filtration can produce beams with the same HVL. The SSD also affects the PDD, and is important for machines which treat at short SSDs.

T183.　　C　　The HVL is a measure of the penetrative quality of a beam. Different combinations of kVp and filtration can lead to beams of the same HVL. HVL can be used to select a PDD table, e.g., from BJR Supplement 17, provided SSD and applicator size are also considered.

T184.　　B　　Using a thinner filter will increase the dose rate, but the energy spectrum will contain more low energy photons.

PD21.

T185.　　E　　Although the isocenter of a plan can be related to internal structures, it must also be related to surface landmarks (preferably triangulation points) in order to position the patient correctly for treatment.

T186. A 1 through 3 are necessary to reproduce the patient's position and the position of the internal organs with respect to skin marks. However, acceptable accuracy can be obtained by tracing around large inhomogeneities such as lungs, bones and air cavities, and assigning bulk densities to these contours.

T187. C Hounsfield numbers range from -1000 for air to about +1000 for bone. The Hounsfield number for water is 0.

T188. D Chest wall thickness and actual lung density can only be reliably determined with a CT taken in the treatment position.

PD22.

T189. A The ratio of treatment times will be the inverse ratio of the TARs for the field size at isocenter, with and without homogeneity corrections. The pathlength to the isocenter without lung corrections is one half the separation (10 cm). The effective pathlength with lung correction will be approximately (7 cm lung) $\times$ (0.3 g/cm^3) + (3 cm tissue) $\times$ (1.0 g/cm^3) = 5.1 cm.

T190. D While air gaps and bone are not generally taken into account for routine head and neck dosimetry with x-rays in the 1-6 MV range, they are much more important for electrons. The anomalous scatter of electrons in inhomogeneous media makes treatment planning more difficult. CT scans are required to localize inhomogeneities, and a pencil beam algorithm should be used for reasonable accuracy in isodose planning.

T191. B As energy increases the % correction for inhomogeneities decreases.

T192. A In a cobalt-60 beam a lung correction of 3.5% per cm is a good approximation.

T193. A Bone decreases the TAR by about 3.5% per cm in a cobalt-60 beam.

T194. C Ignoring lung corrections tends to overdose the prescription point, due to the lower attenuation of lung compared with muscle tissue. Thus it is important to adjust the prescribed dose, depending on whether or not lung corrections are used. The equivalent path length method is an approximation in which the path length $\times$ density is used to calculate beam attenuation (e.g., 10 cm of 0.3 gram/cc lung is equivalent to 3 cm tissue). This does not account for the reduction in scatter from lung to tissue beyond it, and hence tends to slightly overestimate the dose.

PD23.

T195. A A block in the middle of the treatment field reduces the amount of scatter dose more than a block of equal size placed in the corner. Hence field #1 has a larger equivalent square and a higher dose than field #2.

T196.	B	The effect of the blocks is to decrease the area irradiated, and hence the amount of dose due to scattered radiation within the patient. The primary component is a function of the collimator setting, and is essentially unaffected by the presence of the blocks.
T197.	D	The fraction of dose due to scatter increases with both depth and field size.
T198.	D	This consists of separating the dose into primary and scatter, and summing the scatter contributions from small angular segments of the irregular field.
T199.	B	$TAR_{eff} = TAR_0 + SAR$ where TAR_0 is the primary component and SAR is the scatter component for the blocked field. TAR_0 is found by extrapolating TAR(r) to zero field radius.
T200.	A	Since the scatter dose is being estimated for a midline block, it will be proportional to the difference between the SAR for the open field and the SAR for the blocked area. Primary transmission through the block will be on the order of 3-5%, and will be significantly smaller than the scatter dose for cobalt-60.
T201.	B	The effective TAR at $d_{10\ cm}$ is found by adding the TAR_0 at $d_{10\ cm}$ to the effective SAR. The latter is found by dividing the unblocked field into 10° segments centered on the central axis, finding the SAR at d_{10cm} for each radius, and taking the average.
T202.	D	In megavoltage photon beams, the greatest % of scatter is obtained for the lowest energy and greatest depth. Since scatter is removed by secondary blocking the greatest effect on the timer setting will therefore be seen for low energy and large depths.
T203.	E	A thick block will absorb more primary radiation and reduce the dose. Both the width of the block and the amount of unblocked area will affect the quantity of scattered radiation that will reach the midline. The spectrum of the primary radiation will affect both the amount of primary radiation transmitted through the block and the quantity of scattered radiation reaching the midline.
T204.	B	The % transmission by X HVLs = $(100/2^x)$%, therefore 5 HVLs transmit 3%.
T205.	C	5 cm Pb is about 4 HVLs, which would transmit 1/16 or 6.25%. The extra dose is due to scatter from the tissue on either side of the block.
T206.	D	3% is the transmission after 5 half-value layers ($1/2^5$). If 7.5 cm is 5 HVLs, one HVL is 1.5 cm. Thus 6 cm is one less HVL, and the transmission will double from 3% to 6%.

PD24.

T207.	D	By similar triangle geometry: (Field size at extended SSD)/(FS at 100 cm) = (Ext. SSD)/100. Therefore: Ext. SSD = (75/40) × 100 = 187.5 cm.

T208.   A   Block size on tray = Size on film × tray dist./film dist. = $3.0 \times 67/130 = 1.5$ cm.

T209.   B   By similar triangle geometry field size on skin = field size on film × skin distance/film distance = $15 \times 100/130 = 11.5$ cm.

T210.   C   A side (x) of the field at 87 cm can be found by setting up a geometric proportion. x = 10 (87/80) = 10.9 cm.

T211.   B   The field size at 68 cm SSD on the skin, Ws, is given by similar triangle geometry: Ws = 15 (68/80) = 12.8 cm.

T212.   B   The collimator setting (c.s.) is the field size at 80 cm. By similar triangle geometry: c.s. = 40 (80/130) = 24.6 cm.

T213.   B   To keep a field size of 15 cm at 11 cm depth requires a collimator setting C where C = 15 (100/111) = 13.5 cm.

T214.   C   First find the SFD using similar triangle geometry: a 6 cm ring at an SSD of 88 cm projects to 8 cm at SFD: SFD = (8/6) × 88 = 117.3 cm.
Then find the size of block (B) on the tray at 62 cm which projects to 10 cm on the film at 117.3 cm: B = (62/117.3) × 10 = 5.3 cm

T215.   B   Field width (fw) on the skin = (fw on film) × (source skin distance/source film distance) fw = 17.5 × (80-12)/120 = 10 cm

T216.   C   To get a field length of 42 cm on the cobalt unit, which has a maximum field size of 35 cm at 80 cm, the SSD required is (42/35) × 80 = 96 cm.

T217.   A   It is preferable to have a simulator film taken at 80 cm SAD since there are divergence differences between 80 cm SAD and 100 cm SAD. But, if it isn't possible to resimulate, the same film can be used with an "adjusted" TFD. A magnification factor of 1.5 (150/100) is calculated for the 100 cm SAD film. The adjusted TFD for 80 cm SAD is 1.5 × 80 which = 120 cm. A 10 × 10 cm field at 100 cm SAD measures 15 × 15 cm at 150 cm. To have the field size 10 × 10 cm at 80 cm SAD measure 15 × 15 cm at the TFD, the film must be placed at 120 cm TFD.

T218.   E   The distance between BBs on the tray is known and is measured on the film. The proportion of these two numbers is the magnification. Measurement of the distance between the wires on the film divided by the known field size gives the magnification factor. A ring of known diameter can be used on the same principle. A rectangular shaped ruler should not be used (unless it is perpendicular to the beam axis), since it can give a distorted image on the film. A circular or spherical object should be used since the maximum diameter of the circle will always project on the film.

PD25.

T219.   A   The field size at midplane must be kept the same, since this defines the treatment volume. The size on the skin, however, will project smaller, because of the increased divergence at 80 cm SAD.

T220.   D   PDD increases with increasing SSD, hence the increased exit dose. Mayneord's f-factor is the ratio of PDD at the new SSD to the PDD at the original SSD, however, it is typically only a few %. The inverse square correction to the dose rate is by far the most important factor.

T221.   E   The field size at the isocenter must be kept constant to treat the same volume of tissue. However, the field size projected back to the skin will be smaller for 80 cm SAD.

PD26.

T222.   C   Dose-volume histograms give an overall view of the distribution of dose throughout the target volume or critical structure, but cannot give any information about the position of high or low dose regions in the volume. (This information must be obtained from the isodose distribution.)

T223.   E   All of the selections describe measurements which should be part of a comprehensive dosimetry program which would be done before beginning total body irradiation.

T224.   D   A true 3-D system has 3-D description of patient anatomy, 3-D calculations (which include, among other things, accounting for 3-D contours, beam divergence, scatter, block scatter and inhomogeneities), and 3-D matrix of dose points and plan analysis tools like dose-volume histograms. 2-D systems do have inhomogeneity corrections, but assume that the inhomogeneities in the contour extend the length of the fields.

T225.   D   Linear accelerators are now used at many centers in the U.S. and the world for radiosurgery. There are several gamma knives in the U.S. Protons are used at Massachusetts General Hospital and helium ions at Berkeley for Bragg peak radiation for radiosurgery. Electrons are not penetrating enough to be used for radiosurgery.

T226.   B   Field size, SSD and source activity all affect the dose rate of a cobalt-60 unit at d_{max}; the other factors are irrelevant.

T227.   E   An isodose distribution in which 90% of the maximum dose covers the treatment volume requires field widths (at the isocenter) 1.0 cm larger than the volume for most treatment machines. The edge of the field is the 50% decrement line.

THERAPY Electron Dosimetry

ED1. Properties of Electron Beams

T228.   The percent depth dose at 5 cm for a 6 MeV electron beam is approximately:
    A. 100 percent
    B. 90 percent
    C. 80 percent
    D. 50 percent
    E. less than 5 percent

range ≅ 3cm

T229.   What is the approximate range of a 6 MeV electron beam passing through 1 cm tissue overlying the lung (density 0.25 g/cm^3)?
    A. 1 cm
    B. 2 cm
    C. 3 cm
    D. 9 cm
    E. 12 cm

→ Can go 4x as far   *range is:*
1cm tissue + 8cm lung = 9cm total   *range = 6/2 = 3*   *E/2*
8×.25(2)   *b/c ~ 2cm/MeV*
1 + 2 ≅ range ≅ 3cm   *2cm × 4 = 8*   *8+1 = 9*

T230.   The range in cm of an electron beam in tissue is approximately how many times its energy in MeV?
    A. 2
    B. 1
    C. 1/2
    D. 1/4
    E. 1/8

depth of 90% isodose line in cm ≅ E/3

T231.   An electron beam of how many MeV would be most suitable to treat a volume extending to a depth of 5 cm?
    A. 2.5
    B. 10
    C. 20
    D. 5
    E. 15

want to assume 90% isodose line is adequate

T232.   Which of the following properties of electron beams is/are true?
  1. The range in tissue in centimeters is about half the beam energy in MeV
  2. The distance between the 90% and 20% isodose levels on the axis increases with increasing
    energy
  3. The width of the 90% isodose decreases with depth
  4. As energy increases, skin dose decreases
  A. 1, 2, 3
  B. 1, 3
  C. 2, 4
  D. 4 only
  E. all of the above

T233.   The surface dose for a 6 MeV electron beam compared with that of an orthovoltage beam of
  HVL = 0.5 mm Cu is:
  A. higher
  B. lower
  C. the same

T234.   In the treatment of a superficial lesion on the nose, comparing treatment with superficial x-
  rays to treatment with low energy electrons on a linac, the dose to underlying bone will be:
  A. greater for electrons than x-rays
  B. greater for x-rays than electrons
  C. about the same (±3%) for x-rays or electrons

T235.   The reason for treating the parotid with a mixed beam (photons plus electrons) instead of
  electrons alone is:
  A. to reduce the dose to the contralateral parotid
  B. to keep the cord dose below tolerance
  C. to reduce the skin dose
  D. all of the above

T236.   Which of the following is true regarding the surface dose for an electron beam?
  A. depends on the design of the collimators
  B. is the same as that of a photon beam of the same energy
  C. is less for a scattering-foil beam than for a scanning beam
  D. is the same as that of an orthovoltage beam of HVL = 0.5 mm Cu
  E. decreases as energy increases

T237.   The section of the electron depth dose curve labeled A on the diagram is *mainly* due to:
  A. electrons with the highest energy, which have the greatest range
  B. bremsstrahlung x-rays created by electron
    interactions with tissue
  C. characteristic x-rays generated by electrons
    striking the collimators
  D. bremsstrahlung x-rays generated by electron
    interactions with the scattering foil
  E. none of the above

T238.   Compared with that of 6 MeV electrons, a 16 MeV electron depth-dose profile will have:
A. a greater surface dose and a sharper fall off beyond the 90% isodose level
B. a broader plateau region and a lower surface dose
C. a lower bremsstrahlung tail and a lower surface dose
D. a greater distance between the depths of the 90% and 20% isodose levels, and a greater surface dose
E. none of the above

T239.   You can obtain a readable port film for a 20 MeV electron beam head and neck treatment:
A. because there is enough penetration of the 20 MeV electrons through the patient to the film and only a few monitor units are needed
B. same as answer A but the film must be left in for the entire treatment
C. because there is enough bremsstrahlung production and only a few monitor units are needed
D. same as answer C but the film must be left in for the entire treatment
E. under no conditions

T240.   The electron field size required to treat a volume with a 5 cm width at the treatment depth is:
A. 5 cm
B. less than 5 cm
C. wider than 5 cm

isodose curves decrease in width @ greater depths

T241.   To ensure adequate coverage of the treatment volume with an electron beam, it is important to remember that:
A. all isodose curves decrease in width with depth
B. all isodose curves increase in width with depth
C. the 90% isodose increases, and the 20% isodose decreases in width with depth
D. The 90% isodose decreases, and the 20% isodose increases in width with depth

↓ penumbra

T242.   A neck node is to be treated with 8 MeV electrons. Because of the patient's shoulder, an extra 10 cm air gap must be left. Compared with treatment at the nominal SSD, changes due to the air gap would include:
A. an increase in depth dose for the same energy
B. broader penumbra and greater curvature of the beam profile
C. decreased monitor units
D. about twice the amount of bremsstrahlung production
E. skin sparing disappears

↑ distance source
↑ penumbra

T243.   The advantage of a scanned electron beam over one generated with a scattering foil is:
A. higher skin dose
B. lower bremsstrahlung
C. sharper fall off between 90% and 20% depth dose
D. all of the above
E. B and C only

T244.   The position of the secondary collimators on a linac is important when treating with an electron applicator because it can affect:
(A.) output and flatness
B. output only
C. selection of the correct scattering foil
D. ability to attach the applicator
E. selection of the correct energy

ED2. Junctions

T245.   Two 8 MeV electron fields are needed to cover a chest wall. One field is vertical and the other is at a gantry angle of 45°. If the light fields are matched on the skin, the maximum dose at the junction will be:
A. the same as the dose at d_{max} in each field, ±5%
B. about 10% colder than the dose at d_{max} in each field
C. about 10% hotter than the dose at d_{max} in each field
D. about 30% hotter than the dose at d_{max} in each field
E. 50-100% hotter than the dose at d_{max} in each field

T246.   A method of reducing hot spots between adjacent electron fields without introducing cold spots is to:
A. angle the beams towards each other
B. leave a 1.5 cm gap on the skin
C. use different beam energies for each field
D. treat only one electron field per day
E. move the junction two or three times during the course of treatment

ED3. Factors Affecting Output & PDD

T247.   Output for electron cones depends on:
A. position of primary collimators
B. size of cut-out
C. size of cone
D. distance from the source
E. all of the above

T248.   A 4×8 cm rectangular cut-out is placed in a 10×10 cm cone for treatment with 20 MeV electrons. Compared to the measured data for the 10 cm cone, which of the following would be expected to change?
1. depth of the 90% isodose — *less depth for 90% due to scatter*
2. cGy per MU
3. depth of d_{max} (*high x/y small field*)
4. skin dose

A. 1, 2, 3
B. 1, 3
C. 2, 4
D. 4 only
E. all of the above

↓ 50% reduction in field size

ED4. Monitor Unit Calculation

T249.   How much dose is delivered at the 90% PDD level from an electron beam in 100 MU? The output factor is 1.11 cGy/MU.
A. 111 cGy
B. 90 cGy
C. 110 cGy   *$100 = \dfrac{x}{(.90)(1.11)}$*
D. 99 cGy
E. 100 cGy

$100 = \dfrac{x}{(.9)(1.11)}$

$x - 99.9$

T250.   An electron boost is given with a 12 MeV electron beam at 200 cGy per fraction with a 15×15 cm size cone to the 90% isodose line. The machine output is calibrated with a 10×10 cm cone to be 1.00 cGy/MU at d_{max} (100 cm SSD) and the output for the 15×15 cm cone relative to the 10×10 cm cone is 1.02. The MU is:
A. 176 MU
B. 184 MU   *$x = \dfrac{200}{1.02 \, (\cdot) \, (.90)}$*   *$\dfrac{200}{1(1.02)(.9)} = 218$*
C. 218 MU
D. 227 MU   *PDD*
E. 250 MU

T251.   For the previous question, if there is an additional 2 cm air gap from the bottom of the cone to the patient, i.e., the SSD is 102 cm, what adjustment is made to the time-on or MU (assume inverse square holds)?
A. multiply by 0.98
B. multiply by 1.02   *$\left(\dfrac{100}{102}\right)^2 = .96 \quad \frac{1}{.96}$*
C. divide by 1.02
D. multiply by 1.04   *mult $\times$ 1.04*
E. correction is < 1%

ED5. Shielding; Bolus

T252.   A tumor lateral to the spinal cord in the neck is to be treated with a single direct lateral electron field. The deep border of the tumor lies at a depth of 1.5 cm, and the cord is at a depth of 5 cm. Two techniques are considered: (i) 6 MeV electrons prescribed to the 90% isodose level, or (ii) 9 MeV electrons prescribed to the 90% isodose level, using 1 cm of bolus. Which of the following is true?
A. the cord dose will be at least 50% of the prescribed dose, regardless of technique
B. the surface dose will be higher using the 6 MeV technique
C. the dose fall-off beyond the tumor will be steeper with the 6 MeV technique
D. the tumor would be underdosed with the 9 MeV technique

T253.   Lead eyeshields can be used when treating superficial lesions with electron beams. These shields are often coated with wax because:
A. wax is less irritating to the eye than lead
B. wax reduces the bremsstrahlung production in the lead
C. wax reduces the scattered electron dose from the lead
D. all of the above
E. none of the above

T254.   When treating with high energy electron beams one of the problems in using bolus over part of the field is:
A. calculating the thickness necessary
B. finding an appropriate material
C. dose inhomogeneity at the edge of the bolus
D. the production of bremsstrahlung

ED6. Treatment Planning

T255.   A treatment planning computer uses a fan-beam equivalent path length algorithm for electron beam treatment planning. This method is likely to underestimate hot or cold spots in which of the following situations?
1. a single oblique beam, incident at 45°
2. a beam passing through the edge of a region of dense bone
3. a beam passing through a sinus

A. 1 only
B. 2 only
C. 1 and 3
D. 2 and 3
E. 1, 2, and 3

Answers

1.

T228.   **E**   The range of 6 MeV electrons is about 3 cm. Beyond the maximum range of the electrons, the dose is due to bremsstrahlung and is < 5% for 6 MeV electrons.

T229.   **D**   The range in unit density tissue is E/2, i.e., 3 cm. Thus the range will be 1 cm tissue + 2 cm tissue equivalent thickness of lung, or 2/0.25 = 8 cm lung. Thus the total range = 1 + 8 = 9 cm.

T230.   **C**   R (cm) is the practical range of an electron beam. It is the depth at which the falling part of the depth dose curve meets the x-ray (bremsstrahlung) background. R is approximately E(MeV)/2. The depth of the 90% isodose is approximately E(MeV)/3, e.g., a 12 MeV electron beam treats tissue up to about 4 cm depth and spares tissue beyond 6 cm.

T231.   **E**   The "rule of thumb" for electron depth dose is that a beam of A MeV will have the 90% isodose curve at A/3 cm depth.

T232.   **A**   Unlike photons, electron beams have a lower skin dose in lower energy beams.

T233.   **B**   Orthovoltage photon beams have surface doses of approximately 100%, while 6 MeV electrons have some skin sparing.

T234.   **B**   Low energy x-rays deliver a higher dose to bone than tissue because of the Z^3 dependence of the photoelectric effect. In the electron beam absorbed dose is proportional to the mass stopping power, which is similar in bone and tissue.

T235.   **C**   A and B would be better with electrons only. However, the skin dose would be too high so photons are added, and a compromise must be made between the dose to the cord and the skin dose.

T236.   **A**   The surface dose as a % of d_{max} is influenced by the spread of electron energies in the beam and their incident angles. Electrons scattered from the collimator and the foil(s) affect the surface dose. A 6 MeV electron beam typically has a surface dose of about 75 to 95%. A 6 MV photon beam has a much smaller surface dose, typically 20 to 25%, depending on the field size. Scanned electron beams have less low energy scattered electrons compared to electron beams that pass through scattering foils. Orthovoltage beams have surface doses of approximately 100%. Surface dose increases as energy increases.

T237.   D   When electrons interact with high Z materials, bremsstrahlung x-rays and a smaller number of characteristic x-rays are produced. Electron interactions with tissue account for a very small % of the x-ray tail.

T238.   D   As electron energy increases, the distance from the 90% to the 20% isodose levels increases, i.e., the fall-off becomes more gradual, increasing the amount of normal tissue irradiated beyond the tumor. Unlike photons, however, a higher electron energy means a higher surface dose. The bremsstrahlung "tail" depends on a number of factors (such as the scattering foil, applicator design, and loading of the waveguide), but generally it is higher for a higher energy beam.

T239.   D   There are a certain amount of x-rays (bremsstrahlung) produced in the electron mode when the electrons are incident on the scattering foil. It can be on the order of 5% for 20 MeV electrons (for a fairly 'clean' beam). This is enough x-ray contamination to expose a port film, but it will take more than a few monitor units to have a 'readable' port film. The film dose will depend on MU set, patient thickness, and % of bremsstrahlung, but generally the film can be left in place for the entire treatment.

T240.   C   The 80 and 90% isodose curves decrease in width with depth, so the field size on the skin must always be wider than the volume width at depth.

T241.   D   This is important when (a) ensuring adequate coverage of the tumor volume with the 90% isodose curve at depth, (b) avoiding adjacent structures with the penumbra, and (c) abutting electron and x-ray fields.

T242.   B   The greatest effect on increasing SSD for electrons is a broadening of the penumbra, and a rounding of the beam profiles. Thus, although the light field appears larger, coverage by the 90% isodose will not be the same % of the light field. The depth dose curve is essentially unchanged but monitor units will be increased to account for decreased output.

T243.   E   A scanned electron beam achieves a large flat beam without passing through a scattering foil, and so has less bremsstrahlung contamination. Interactions in the foil also tend to spread the energy spectrum. The scanned beam therefore has a narrower spectrum leading to a lower skin dose, broader plateau, and sharper fall off.

T244.   A   Scatter from the secondary collimators, or "jaws," affects the field flatness, and must be set to some predetermined setting when the electron cone is attached. The output can also be affected by the collimator setting.

2.

T245.   E   The actual dose depends on the edge characteristics of the isodose curves, but angling electron fields towards each other always results in hot spots; other measures such as moving the junction, and/or adding compensators may reduce the value of the hot spot.

T246.   E   Although this will not eliminate a hot spot, it will reduce, for instance, a 30% hot spot to a larger area of 10% if the junction is moved three times. Electron fields should never be angled towards each other, as this can easily result in a complete overlap (200%). If gaps are left on the skin, a cold spot will result.

3.

T247.   E   Output for electron beams depends on all the factors listed.

T248.   E   Care should be taken when prescribing high energy electrons with small fields. The depth of the 90% isodose can be significantly less, due to lack of scatter from the blocked area. If the depth is adequate, the output for the small insert will need to be measured, and the skin and build up characteristics should be expected to change due to scatter from the insert.

4.

T249.   E $MU = dose / (output \times PDD)$
$100 = dose / (1.11 \times 0.90)$ dose = 100 cGy.

T250.   C $\dfrac{200 \text{ cGy}}{0.9 \times 1.02} = 218 \text{ MU}$

T251.   D   If the patient is farther away by 2 cm, the output will be lower by the inverse square law, $(100/102)^2$. The original time-on should be increased by 1.04. Tables of measured % correction per cm air gap can also be used.

5.

T252.   C   The dose fall-off beyond the tumor will be steeper with the lower energy technique. The rule of thumb for electron range is E/2 cm. If the cord lies beyond the range for both beams, it is expected to get less than 10% of the prescribed dose. The 9 MeV technique will have a higher surface dose because of the bolus. The rule of thumb for the depth of the 80-90% isodose is E/3 cm. Therefore the tumor would be adequately treated with either technique.

T253.   C   Electrons are scattered from the lead and absorbed in the wax. Fewer scattered electrons are created in the wax because scatter is strongly Z-dependent.

T254.   C   If there is a sudden discontinuity in the bolus, there may be hot and cold areas in the dose distribution in tissue below the discontinuity.

6.

T255.   E   Situations in which increase or decrease in scatter occurs are not well modeled by this algorithm. With inhomogeneities or sloping surfaces, it is very important to be able to take account of scatter in calculating an electron beam dose distribution. A pencil-beam algorithm is generally better at modeling these situations; however, any algorithm must be thoroughly evaluated by comparison with measured data, before reliance can be placed on treatment plans generated.

THERAPY Brachy-therapy

B1. Decay of Activity

T256.　A cesium-137 source has an activity of 15.5 mg-Ra eq. in June 1980. Its activity in Dec. 1986 was _____ mg-Ra eq. [$T_{1/2}$ = 30 y; Γ = 3.3 (R-cm^2)/(mCi-hr); HVL = 0.6 cm Pb]

A. 12.5
B. 8.78
C. 13.3
D. 14.7
E. 17.8

$$e^{\frac{-.693(6.5)}{30}} = .8606 \quad A = A_0 e^{\frac{-.693 t}{T_{1/2}}}$$

$$15.5(.8606) = 13.34$$

$T_{1/2} = 5.26y$

T257.　A cobalt-60 source has a certificate stating that the exposure rate at 1 meter is 120 R/min on Jan 15. If the source is installed on June 15, the dose rate to a small mass of tissue *in air at 80 cm* on that date is _____ cGy/min.

A. 151
B. 85
C. 70
D. 170
E. 115

$$\frac{-.693(4.17)}{5.26}$$

$$e^{\frac{-.693(.417)}{5.26}} = .947$$

$T_{1/2} = 5.26 y$ 5mo/2mo = .417

$(.947)(120)\left(\frac{100^2}{80}\right)(.957)$

$120(.951)$　$= 170$

f electr soft tissue = .95

Abs dose = f X

T258.　An isotope with a half-life of 60 days has an initial activity of 10 mCi. After 30 days its activity will be _____ mCi:

A. 5.0
B. 6.1
C. 7.1
D. 7.5
E. 7.9

$$= 10 \left(\frac{1}{2}\right)^{.5}$$

$$10 e^{-.693(.5)} = 7.07$$

$$10 e^{-.693\left(\frac{30}{60}\right)}$$

$$10e^{-.693 \frac{30}{60}}$$

T259.　A cobalt-60 source with a dose rate of 200 cGy/min was installed on Jan. 1. On Nov. 1, the dose rate will be approximately _____ cGy/min:

A. 198.1
B. 179.2
C. 163.5
D. 100.0
E. 31.7

$$\frac{10 \, mo}{12 \, mo} = .833$$

$$200 e^{\frac{-.693(.833)}{5.7}} = 180.7$$

$$= 200 \, e^{-.693(.833)}$$

T260. The half-life of iodine-125 is 60 days. In 6 months the activity will be ____ its original value:
A. in transient equilibrium with
B. 50% of
C. 1/8 of
D. 5% of
E. 1/64 of

(handwritten:) $t_{1/2} = 2\,mo$ $6\,mo = 3\;t_{1/2}$

$e^{-\frac{.693\,(183)}{60}} = .12 \approx \frac{1}{8}$

$\left(\frac{1}{2}\right)^3 = \frac{1}{8}$

T261. The half-life of iridium-192 is 74 days. The decay constant is:
A. 3.7 days
B. 37 days
C. 106.8 days
D. 0.0094 per day
E. 0.027 per day

(handwritten:) $\lambda = \frac{.693}{74} = .0094/day$

$\lambda = \frac{.693}{t_{1/2}}$

T262. A newly installed cobalt-60 teletherapy source is found to have an output of 210 cGy/min at the isocenter. If the licensee does **not** want the dose rate to fall below 100 cGy/min, when should the source be replaced?
A. 3 yrs 11 months
B. 5 yrs 3 months
C. 5 yrs 7 months
D. 6 yrs 1 month
E. 6 yrs 6 months

(handwritten:)
$100 = 210\, e^{-.693\,t/(5.7)}$ take ln of both sides
$.476 = e^{-\frac{.693t}{5.7}}$

$100 = 210\, e^{-\frac{.693\,t}{5.7\,y}}$ $.742 = \frac{.693t}{5.7}$

B2. Dose Rates Around Line & Point Sources

(handwritten:) $t = 6.102$ 6 yrs 1 month

T263. For a Fletcher applicator with loading of 15-10-10 mg-Ra in the tandem and 15 mg in each ovoid, a lateral displacement of 0.2 cm in the location of point A would cause about a _____ % change in the dose rate at this point.
A. 0.1
B. 1
C. 2
D. 10
E. 30

(handwritten:) close to source $\sim \frac{1}{r}$ $\frac{1}{2.2} = .45$ $y = .5$ 10% dcf

no only for Pts far away $\left(\frac{2}{2.2}\right)^2 = .83$ not quite 17% dc   not all sources are 2 cm away $1000\,mR \approx 1R$

T264. The exposure rate at 1 m from 100 mg of radium is:
A. 8.25 R/hr
B. 8.25 mR/hr
C. 82.5 mR/hr
D. 82.5 mR/min
E. 0.825 R/hr

(handwritten:) for Ra mg = mCi

$\Gamma = 8.25 \left(\frac{R\,cm^2}{mCi\text{-}hr}\right)$

$\frac{100 \, (8.25)}{(100^2)}$ $(100\,mCi)(8.25\,\frac{R\,cm^2}{mCi\text{-}hr})\left(\frac{1}{100\,cm}\right)^2$

$= .0825 \; R/hr$ $82.5 \; mR/hr$

$\Gamma_{ra} = 8.25 \frac{R\,cm^2}{mCi\text{-}hr}$

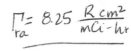

T265. The exposure rate 4 meters from a patient containing 60 mg-Ra eq. filtered with 0.5 mm Pt is about ____ mR/hr. Consider the applicator to be a point source and that the patient absorbs about 30% of the radiation.

A. 1
B. 2
C. 3
D. 4.5
E. 16

$1 - 30\%$

$8.25(60)\left(\frac{1}{4m}\right)^2 = 30.9\,(.7)(100)^2$

$= 2.16 \frac{mR}{hr}$

$\Gamma \frac{1}{r^2}$

$\left(8.25\frac{R\,cm}{mCi\,hr}\right)\left(\frac{1}{400cm}\right)^2 = 30.9 \times .7 = 2.16$

T266. 164 mg-Ra eq. of iridium seeds are placed in a shipping container with a diameter of 30 cm. In order to reduce the exposure rate on the surface of the container to less than 6 mR/hr the number of tenth value layers (TVLs) of lead surrounding the seeds must be:

A. 3
B. 6
C. 10
D. 100
E. 1000

$8.25(164)\left(\frac{1}{15}\right)^2 = 6013 \frac{mR}{hr}$

$6013(.1)(.1)(.1) \cong 6$

$A \cdot \Gamma \cdot \frac{1}{r^2}\; 164\,(8.25)\left(\frac{1}{15}\right)^2 = 6013$

T267. The rapid fall-off of dose around a cesium-137 source in a 5 cm radius of tissue is due primarily to:

$N_9 = .67$

A. the inverse square law
B. tissue attenuation → compensated by scatter
C. alpha and beta emission — Cs does not emit α's and β's are absorbed in source encapsulation
D. all of the above
E. none of the above

T268. Three 20 mg-Ra sources of physical length 2 cm and active length 1.5 cm are loaded into a vaginal cylinder of diameter 3 cm. Using the table below, the dose rate on the surface of the cylinder opposite the center of the second source is ____ cGy/hr:

Perpendicular Dist. From Source (cm)	cGy per mg-hr in Tissue for Active Length 1.5 cm				
	Distance Along Source Axis (cm)				
	0	1	1.5	2	3
1	6.1	4.2	2.7	1.7	0.7
1.5	3.0	2.4	1.8	1.3	0.7
2	1.8	1.5	1.2	1.0	0.6
3	0.8	0.8	0.7	0.6	0.4

A. 112
B. 162
C. 40
D. 76
E. 122

$A^2 + b^2 = c^2$

$.25 + 4 = 4.25\;\;\sqrt{4.25}=$

$= 3\frac{cGy}{mg\,hr}(20) + 1.3(20) + 1.3(20)$

$= 112\,cGy/hr$

$\left(3\times 89\%/mhr \times 20\,mg\,Ra\right) + 1.3(20)\times 2$

$60 - 52 = 112$

Therapy ▲ Brachytherapy Questions ▲ Page 181

I-125
T½ = 60 days

$\Gamma_{CS} = 3.3 \frac{R cm^2}{mCi \cdot hr}$ or $3.26 \frac{R cm^2}{mCi \cdot hr}$

T269. An "anisotropy correction" is used in treatment planning to correct for:
A. differential attenuation of an electron beam by tissues of different densities
B. the normalization of a treatment plan with multiple beams having more than one isocenter
C. the decay of iridium-192 seeds during a temporary implant
D. changes in the isodose pattern due to oblique incidence of photon beams
E. increased absorption causing a reduction in dose rate along the axis of an iodine-125 seed

isotropic

from titanium encapsulation

T270. To calculate the exposure rate from a point source high energy gamma emitter, all of the following are required *except:*
A. current activity
B. distance from the source
C. specific exposure-rate constant
D. mean life

T271. The PDD in tissue at a given depth from a vaginal cylinder will be greatest for the _____ applicator:
A. largest diameter
B. smallest diameter
C. depends on the depth

reduces $\frac{1}{r^2}$

can use $\Gamma_{Ra} \Gamma$

T272. A cesium-137 source carrier contains 155 mg-Ra eq. How many HVLs of lead are required to reduce the exposure level at 1 meter to 2 mR/hr?
A. 2
B. 6
C. 8
D. 10
E. 14

$155 \times 8.25 \frac{1}{(100)^2} = 127.9$
64
32
16
8
4

$\Gamma_{CS} = 8.25 \frac{R cm^2}{mCi \cdot hr} (155)(\frac{1}{100})^2 =$

$127.9 \frac{mR}{hr}$

6 HVLs go to $1.998 \frac{mR}{hr}$

B3. Total Dose From Permanent Implants

T273. A permanent implant is performed with iodine-125 seeds, with a half-life of 60 days. The total dose at a point is 18,000 cGy. The initial dose rate is _____ cGy/hr.
A. 8.7
B. 12.4
C. 18.0
D. 208.8
E. 150.0

mean life = $T_{1/2}(1.44)$ = 2073.6 hrs
total dose = initial dose rate × mean life.

$18000 = X_i \times 2073.6$

$X_i = 8.68$

T274. An iodine-125 seed implant delivers an initial dose rate of 10 cGy/hr. The half-life of iodine-125 is 60 days. The total dose delivered by the implant is _____cGy.
A. 20736
B. 864
C. 10000
D. 417
E. 15403

mean life = 1.44 T½
(2073.6 hrs)(10 cGy/hr) = 20,736 cGy
1.44(1440)(60) = 2073
× 10 =

T275.   The average life of a radioisotope is:
          A. used to calculate the total dose delivered by a permanent seed implant
          B. equal to 1.44 / half-life
          C. used to calculate the dose rate mid-way through a temporary insertion
          D. also called the decay constant
          E. all of the above

B4. Units

T276.   An isodose distribution is plotted for a ^{198}Au implant, and the target volume is covered by the
          "50" isodose level. However, it is not stated whether this represents "initial cGy/hr" or "dose
          to total decay, in Gy." The half-life of ^{198}Au is 2.7 days. Pending further clarification, what can
          be said about the uncertainty in dose delivered?
          A. there is no difference, as "initial cGy/hr" = "dose to total decay, in Gy"
          B. by assuming "initial cGy/hr," the patient could receive 7% more dose, if the "50" was
             actually total Gy
          C. by assuming "Gy to total decay," the patient could get 35% less dose, if the "50" was
             actually "initial cGy/hr"
          D. not enough information to determine

T277.   The number of Bq in one Ci is:
          A. 1
          B. 10^7
          C. 3.7×10^7
          D. 3.7×10^{10}
          E. 2.7×10^{-11}

T278.   1 Bq is:
          A. 3.7×10^{10} Ci
          B. 3.7×10^7 Ci
          C. 2.7×10^{-11} Ci
          D. 2.7×10^{-8}
          E. none of the above

T279.   Cesium-137 activity is expressed in terms of mg-Ra eq. because:
          A. the activity in millicuries is difficult to measure accurately
          B. the gamma-ray energy is the same
          C. Patterson-Parker tables designed for radium can be used without modification
          D. shielding requirements are the same for 1 mg radium and 1 mg-Ra eq. cesium-137
          E. all of the above

T280.   One mg-Ra equivalent of cesium-137:
          A. has the same activity as one mg of Ra
          B. has an activity of one mCi of cesium-137
          C. has a higher exposure rate at 1 m than 1 mg of Ra
          D. is equal to 8.25 mCi of cesium-137
          E. is equal to (8.25/3.26) mCi of cesium-137

Ir-192 $\boxed{\Gamma = 4.6}$

T281. A radiation oncology department is finally changing from radium sources to cesium-137 sources for brachytherapy. A 10 mg radium source should be replaced by a _____ cesium-137 source (ignoring filtration). [exposure rate constant for ^{137}Cs = 3.26 (R-cm²/mCi-hr) at 1 cm]

A. 25 mCi
B. 10 mCi
C. 9 mCi $\frac{8.25}{3.26}(10) = 25$
D. 10 mCi
E. 0.25 mCi

T282. A vaginal cylinder is loaded with 20-15-20 mg-Ra equivalent sources of cesium-137. What is the total number of mCi of cesium-137? [exposure rate constant for ^{137}Cs = 3.26 (R-cm²/mCi-hr) at 1 cm]

A. 16.6
B. 21.7 $\frac{8.25}{3.26}(55) = 139$
C. 55.0
D. 139.2
E. 352.2 $\underbrace{8.25\,(55)}_{Radium} = \underbrace{3.26\,(mCi)}_{Cs}$

T283. After loading an iridium-192 implant, it is discovered that the sources were each 0.5 mg-Ra eq., instead of 0.5 mCi as intended. The consequence is:

A. the dose rate in the patient will be higher
B. the dose rate in the patient will be lower
C. the dose rate will be the same, as mg-Ra eq. = mCi for this isotope

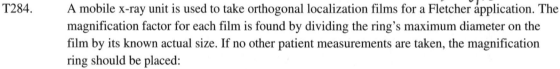

B5. 3-D Coordinate Reconstruction

T284. A mobile x-ray unit is used to take orthogonal localization films for a Fletcher application. The magnification factor for each film is found by dividing the ring's maximum diameter on the film by its known actual size. If no other patient measurements are taken, the magnification ring should be placed:

A. on the patient's skin, closest to the x-ray tube, and at the center of the field for both films
B. on the patient's skin, closest to the film cassette for both films
C. at the patient's midline for the lateral film and at the level of the applicator for the AP film
D. on the patient's anterior skin at midline for both films
E. none of the above

T285. A small radioactive sealed source is located at (0, 0, 2) in a three dimensional (cm³) coordinate system. A critical organ is situated at (-2, 3, 4). The distance between them is _____ cm.

A. 1.7
B. 3.0
C. 4.1 $\sqrt{(-2)^2 + 3^2 + 2^2} = 4+9+4$
D. 5.0
E. 7.0 $= \sqrt{17} = 4.1$

T286.	Which of the following could be used to reconstruct the 3-dimensional positions of a set of implanted sources for dosimetry purposes?
	A. shift films	→ what are these?
	B. AP and lateral films
	C. oblique orthogonal films
	D. all of the above
	E. A and B only

B6. Systems of Implant Calculation

T287.	Using Patterson-Parker tables, a total of 3300 mg-hrs are required for a 4×4 cm single plane applicator with a treating distance of 1 cm. Ra needles are available with activities of 2.0, 3.0, and 4.0 mg, and active lengths of 4.0 cm. The treatment time would be:
	A. 237.5 hrs
	B. 206.3 hrs
	C. 183.3 hrs
	D. 165.0 hrs
	E. not enough information given

T288.	Patterson-Parker tables can be used to calculate treatment times for all of the following *except*:
	A. a radium tube mould
	B. a cesium-137 needle volume implant of the tongue
	C. a needle implant without crossed ends
	D. an implant with equal activity per cm around the periphery
	E. an iridium-192 breast implant with equal activity seeds in lines 1 cm apart

T289-292.	The latest modifications of the Patterson-Parker Rules for radium implants and surface applicators (answer A for true and B for false):

T289.	Require a non-uniform distribution of radium for a uniform dose distribution.

T290.	Require that the milligram-hours (mg-hr) be increased as the spacing between a two-plane implant is increased.

T291.	Require corrections for oblique filtration caused by the platinum-iridium walls of the radium tubes and needles.

T292.	Take into account the radiobiological effectiveness (RBE) of radium gamma-rays.

Therapy ▲ *Brachytherapy Questions* ▲ *Page 185*

T293. A two plane interstitial radium needle implant is planned using the Patterson-Parker System under the following conditions.

Area of each plane: 20 cm^2
Separation between planes: 1 cm *[handwritten: ≥/ cares → 0.5cm, to distance]*
Total time: 6 days
Total dose: 6000 cGy

	mg-hr/1000 cGy treatment distance	
Area (cm^2)	0.5 cm	1.0 cm
10	230	430
20	(370)	640
30	490	800
40	600	930

[handwritten: each plane will deliver 1/2 the dose → 370 mg hr (3000 cGy) / (6x24) hrs (1000 cGy) = 7.7 mg]

Approximately how many mg-Ra should be used in **each** plane?
A. 7.7
B. 13.3
C. 15.4
D. 26.7
E. 27.0

[handwritten: total = 15.4]

T294. A radium-226 treatment volume crossed at both ends is planned to deliver 6000 cGy in 6 days. However, in the operating room only one end of the implant is crossed. For calculation purposes, the _____ is reduced by 7.5%.
A. dose rate
B. total activity
C. treatment time
D. implanted volume
E. mg-hr/1000 cGy

[handwritten: if both end uncrossed ↓ in volume by 15%]

T295-297. Match the implant with the description:
A. uniform spacing of equal strength sources, and use of a "nomograph" to determine source activity
B. nonuniform distribution of sources; more activity on the periphery
C. prescribe to 85% of the average dose

T295. B Patterson-Parker *[handwritten: (Manchester)]*

T296. C Paris *[handwritten: for Ir-192 plane wires]*

T297. A Memorial Hospital/New York *[handwritten: (Quimby (?)) for approved]*

T298.　　Which of the following is not a feature of the Patterson-Parker (Manchester) implant system?
　　　　A. non-uniform source loading
　　　　B. reducing equivalent area by 10% for each uncrossed end
　　　　C. dose uniformity to ± 3% in treatment plane, if the rules are followed　*only to ± 10%*
　　　　D. linear sources with 2/3 loading in the centers of large rectangles
　　　　E. maximum allowed distance between active ends of sources depends on treating distance chosen

[handwritten: ?]

[handwritten table/notes in left margin:]
< 25 / 25-100 / >100
chr ½ / ½ / ⅔
mgys ⅓ / ½ / ⅓

B7. Advantages of Cesium-137 Over Radium

T299.　　The following are advantages of using cesium-137 sources over radium sources, *except*:
　　　　A. lower maximum photon energy
　　　　B. smaller HVL in Pb
　　　　C. shorter half-life　*not an adv. have to buy new sources*
　　　　D. less exposure to personnel
　　　　E. less hazardous if broken

T300.　　Cesium-137 sources used for implants have a similar isodose distribution to radium sources because:
　　　　A. they have similar half-lives
　　　　B. they have similar gamma energies
　　　　C. the inverse square law dominates the distribution　*— for all radium isotopes*
　　　　D. the same activities can be used
　　　　E. all of the above

[handwritten right: Γ = 8.25]
< ¹³⁷ Cs Γ 3.26
¹⁹²I Γ 4.6
radon Γ 8.05

T301.　　The major reason cesium-137 sources have replaced radium-226 sources for implants is:
　　　　A. cesium-137 has better depth dose characteristics
　　　　B. cesium-137 emits monoenergetic gamma rays
　　　　C. the exposure rate constant is less for cesium-137 than for radium-226　*} true, but irrevelant*
　　　　D. cesium-137 requires less shielding and is less hazardous　*→ blc radium can leak radon gas.*
　　　　E. cesium-137 is easier to manufacture than radium-226

B8. Shielding

[handwritten: 485 8.25 / (1000)² = .004 R/hr = 4mR/hr]
[handwritten: 5 mSv = .5 rem]

T302.　　A safe contains 485 mg Ra-eq of Cs-137. A secretary has a desk in an office 10 meters away from the safe. Neglecting the office wall, how many HVLs of Pb are needed in the wall of the safe to reduce the annual dose equivalent to 5 mSv?
　　　　A. 1
　　　　B. 4
　　　　C. 7
　　　　D. 11
　　　　E. 15

[handwritten: 5mSv = .005 Sv = .5 rem/yr = 500 mrem
for 40 hr work week = .25 mrem/yr]

[handwritten: 485 × 8.25 / (1000)² = .004 R/hr = 4 mR/hr (× 52 Wx 40 h) = 8320 mR/yr]

[handwritten: from 4 to .25 = 16 = 2⁴]
[handwritten: 1 cGy = 1 rad = 100 mR / 1 rem = .01 Jv / 1 Sv = 01 rem = 10 mSv]

T303.   Which of the following would **most** reduce exposure to family members visiting a patient with a cesium needle implant?
A. reducing the visiting time by one half
(B) doubling the distance between the patient and the visitors
C. placing a Pb screen of thickness one HVL between the patient and the visitors
D. none of the above, since all have an equal effect

T304.   A 2 cm thick lead shield is used to reduce the dose rate from a radioactive implant. If the HVL for this radiation is 5 mm, the dose rate will be reduced by a factor of:
A. 2
B. 4
C. 10
(D) 16
E. 100

$2^4 = 16$

$\frac{2.0}{.5} = 4 \, HVL$

B9. Properties & Use of Clinical Radionuclides

T305.   Iodine-125 seeds are used in brachytherapy for their _____ radiation.
A. auger electron
B. positron
(C) x-ray and gamma ray
D. alpha particle
E. beta minus

$\sim 27-35 \, KeV$

$I^{125} \, r \, Pd^{103} \, mostly \, EC$

T306. ^{32}P decays by beta emission. The transition energy is stated as 1.7 MeV. This means that:
A. the average beta energy is 1.7 MeV
B. the most probable beta energy is 1.7 MeV
(C) the maximum beta energy is 1.7 MeV
D. the minimum beta energy is 1.7 MeV
E. all the betas are emitted with an energy of 1.7 MeV

T307.   Cesium-137 decays by:
(A) beta followed by gamma
B. gamma followed by alpha and beta
C. alpha followed by gamma
D. photodisintegration
E. none of the above

γ is the metastable state of ^{137m}Ba

constant - gamma due to decay (not B!)

T308.   The exposure-rate constant for cobalt-60 (Γ_{Co}) is approximately four times greater than the exposure rate constant for cesium-137 (Γ_{Cs}), although the energy is only twice as high. This is because:
A. ^{60}Co has more energetic beta-rays
B. lower energy ^{137}Cs photons are more likely to be absorbed by the photoelectric process
C. ^{60}Co has a lower atomic number (Z)
(D) ^{60}Co emits two photons per disintegration
E. ^{137}Cs emits only one photon with a fractional abundance (n) of about 0.5

T309.	The useful gamma-ray emissions from a radium source in equilibrium with its daughter products come from:
A. platinum-196
B. radon-222
C. radium-226
D. uranium-238
E. none of the above

T310-317.	Match the radioisotope with its clinical use and properties (answers can be used more than once):
A. radium-226
B. phosphorus-32
C. iodine-125
D. iridium-192
E. cesium-137

T310.	C	Can be used for permanent seed implants.

T311.	D	Has a half-life of 74 days.

T312.	A	Emits alphas.

T313.	B	Is a pure beta emitter.

T314.	E	Has a gamma energy of 0.66 MeV.

T315.	C	Emits 27 and 35 keV photons.

T316.	A	Has the greatest HVL in Pb.

T317.	E	Has a half-life of 30 yrs.

T318.	The radioisotope used to treat polycythemia vera is:
A. phosphorus-32
B. iridium-192
C. iodine-125
D. iodine-131
E. sodium-22

T319-321.	Match the time for a 1% reduction in activity to the appropriate radioisotope.
A. cobalt-60
B. cesium-137
C. iodine-125
D. radium-226

$.99 = Ae^{-.693 \, t/t_{1/2}}$

T319.	C	1 day
T320.	B	5 months
T321.	A	1 month

T322.   Cesium-137 is:
            A. created by bombarding cesium-138 with neutrons
            B. created by bombarding a target with deuterons in a cyclotron
            C. a fission product, separated from used reactor fuel rods
            D. a naturally occurring radioisotope found in uranium ore
            E. none of the above

T323-326.   Match the radioactive source with the approximate decay rate:
            A. <1% per year
            B. 2.3% per year
            C. 1% per month
            D. 5% per day
            E. 1% per day

$\mu = \frac{.693}{T_{1/2}} \times 100\% = \div 12\,mo$

T323.   A   Radium-226 $T_{1/2} = 1600\ yrs$

T324.   C   Cobalt-60 $T_{1/2} = 5.26\ yrs$

T325.   B   Cesium-137 $T_{1/2} = 30\ yrs$

T326.   E   Iridium-192 $T_{1/2} = $ days 74.2

T327-331.   Match the photon energy with the radionuclide:

T327.   D ^{137}Cs   A. 0.030 MeV avg
T328.   B ^{99m}Tc   B. 0.141 MeV (laws for energy - monochromatic)
T329.   A ^{125}I   C. 0.38 MeV avg
T330.   C ^{192}Ir   D. 0.662 MeV monoenergetic Cs
T331.   E ^{60}Co   E. 1.25 MeV avg 1.17 & 1.33

T332.   Iodine-125 seeds are a good alternative to iridium-192 for temporary brain implants because:
            A. it is easier to determine depth dose for iodine-125 than for iridium-192
            B. it has a lower RBE than iridium-192   low Nf gives higher RBE due to low Ng e emitted
            C. it has a longer half-life than iridium-192
            D. it has a lower photon energy which gives less exposure to medical personnel
            E. the exposure rate constant is higher than iridium-192

T333-336.   Select the property most appropriate to each isotope:
            A. is a beta emitter used in liquid form
            B. emits gamma rays of energy about 35.5 keV
            C. has a half-life of 17 days
            D. is a member of the uranium decay chain mass # > 200
            E. is in secular equilibrium with its beta-emitting daughter 90Y

T333.   B ^{125}I
T334.   E ^{90}Sr
T335.   A ^{32}P
T336.   C ^{103}Pd

B10. Strontium Eye Applicators

T337.   Regarding strontium eye applicators, which of the following is *false*?
   A. the betas from ^{90}Sr have an energy of 540 keV
   B. the betas from ^{90}Y have an energy of 2.27 MeV
   C. the lens dose for an eye applicator is about 20%
   D. the half-life of ^{90}Sr is about 28 yrs
   E. the foil covering the strontium is to scatter the ^{90}Sr betas, and improve dose uniformity
      across the eye

   doesn't achieve better
   only the β's from ^{90}Y treat the eye, the β's from ^{90}Sr are absorbed in the shield

T338.   The emissions from a strontium-90 source used in radiation therapy are:
   A. alpha, beta and gamma
   B. gamma only
   C. alpha only
   D. beta only
   E. positrons

T339.   Strontium-90 eye applicators use _____ max beta-rays from _____ for treatment
   purposes.
   A. 0.54 MeV ....^{90}Sr
   B. 2.27 MeV ....^{90}Zr
   C. 0.54 MeV ....^{90}Y
   D. 2.27 MeV ....^{90}Y
   E. 2.27 MeV ....^{90}Sr

T340.   Which of the following is true of strontium-90?
   A. it is a pure beta emitter
   B. it is in equilibrium with its radioactive daughter
   C. it is used to make eye applicators
   D. it has a half-life of 28 years
   E. all of the above

B11. Equilibrium

T341.   A radium source _____ with its daughters.
   A. is in transient equilibrium
   B. is in secular equilibrium
   C. never achieves equilibrium

T342.   When secular equilibrium is achieved:
   A. the parent decays with the half-life of the daughter
   B. the daughter decays with its own half-life
   C. the daughter decays with the half-life of the parent
   D. the rate of decay of the daughter is determined by the initial activity of the parent
   E. none of the above

   $T_{1/2 d} < T_{1/2 p}$

   apparent
   ½ life

T343-345. Match the type of equilibrium with its decay:
           A. secular
           B. transient

T343. A $^{226}Ra \rightarrow ^{222}Rn$

T344. B $^{99}Mo \rightarrow ^{99m}Tc$

T345. A $^{90}Sr \rightarrow ^{90}Y$ T½ slow
       T½=28y   T½=slow

B12. Gynecological Applicators

T346.   The advantage of using <u>larger diameter</u> ovoids in a Fletcher-Suit applicator is:
           A. the large ovoids are easier to see on a localization radiograph
           B. the larger the ovoid the higher the mucosal dose rate
           C. a lower mucosal dose and better depth dose distribution   less fall off,
           D. a lower dose to the bladder and rectum   remove us!
                                                                         r²

B13. Miscellaneous

T347.   After removing iridium seeds from a patient one should always:
           A. call the radiation safety officer to perform a room survey within the next 24 hrs
           B. monitor the patient and bedding for remaining sources with a sensitive detector
           C. order radiographs to detect any remaining sources
           D. count each of the sources to verify that they are all present
           E. none of the above are necessary

T348.   Which of the following is the most important advantage of brachytherapy over teletherapy?
           A. there is no repair of sublethal injury
           B. there is a more homogeneous dose delivered
           C. the volume of normal tissue treated is minimized
           D. the oxygen enhancement ratio is reduced
           E. hazards to personnel are less

Answers

B1.

T256.   C   Activity = $A_o \times \exp[(-0.693) \times (t)/T_{1/2}] = 15.5 \exp[(-0.693 \times 6.5)/30)]$
= 13.3 mg-Ra eq.

T257.   D   The exposure rate at 1 meter after 5 months decay is:
$120 \times \exp([-0.693 \times 5]/[5.26 \times 12]) = 113.6$ R/min. The dose rate at 80 cm is $113.6 \times (100/80)^2 \times 0.957 = 169.9$ cGy/min.

T258.   C $A_t = A_o \exp - (0.693 \times 30/60) = A_o \times 0.707 = 7.07$ mCi.

T259.   B   Cobalt-60 decays about 1% a month and the time elapsed between Jan. 1 to Nov. 1 is 10 months $(0.9 \times 200 = 180)$. The exact answer is dose rate (Nov) = dose rate (Jan) $\exp - [(0.693 \times 10/5.26 \times 12)] = 179.2$ cGy/min.

T260.   C   It will decay by a factor of $(1/2^3) = 1/8$, since 6 months is 3 half-lives.

T261.   D   Decay constant = $(\ln 2)/T_{1/2} = 0.693/74 = 0.0094$ day^{-1}.

T262.   C   The half-life of cobalt-60 is 5.26 yrs.
Final dose rate = Initial dose rate $\times \exp^{-(0.693 \times t/T_{1/2})}$
$100 = 210 \times \exp^{-(0.693 \times t/5.26)}$
$t = 5.63$ yrs

B2.

T263.   D   This can be calculated roughly using the inverse square law: $(2/2.2)^2 = 0.83$, i.e., 17%. If all sources were 2.0 cm from point A, this would be the exact answer; however, since most sources are further away, the answer is about 10%. In some computer systems it is difficult to localize point A better than ± 2 mm, so this should be borne in mind when considering the accuracy of the calculated dose rate at point A, especially when checking the calculation by hand.

T264.   C   Exposure rate = Exposure rate const $\times$ activity $\times (1/r^2)$
= 8.25 (R-cm^2/mg-hr) $\times 100$ mg $\times (1/10,000$ cm$^2) = 82.5$ mR/hr.

T265.   B $R = (\Gamma \times A)/d^2$ where Γ is the exposure rate constant, A is the activity and d is the distance from the source. The exposure rate is reduced by 0.7 by the patient. R = 8.25 R-cm^2/(mg-hr) $\times 60$ mg $\times 1000$ (mR/R) $\times 0.7/(400$cm$)^2 = 2$ mR/h.

T266. A Assume a point source of 164 mg-Ra eq. at the center of the container. Exposure rate = (Activity (mg-Ra eq.) $\times \Gamma_{Ra}$) / d^2 = (164×8.25) $(1/15^2)$ = 6 R/h. To reduce 6 R/h to 6 mR/h, a reduction of 10^{-3} is required, or three 10th value layers.

T267. A The inverse square law has a marked effect at short distances, e.g., between 1 cm and 2 cm the dose rate falls by a factor of $(1/2)^2 = 0.25$. Attenuation and scatter approximately balance within the first 5 cm. The energy would have to be much lower to cause appreciable attenuation in this radius. Cesium does not emit alphas, and the betas are absorbed in the source encapsulation and do not contribute to the dose around the source.

T268. A Perpendicular distance (for each source) = 1.5 cm. Longitudinal distance from center of source to calculation point = 2 cm, 0 cm, 2 cm respectively. Total dose rate = sum of (cGy per mg-hr $\times$ mg-Ra) for all sources = $(1.3 \times 20) + (3 \times 20) + (1.3 \times 20)$ = 112 cGy/hr.

T269. E Increased absorption in the welds of the titanium encapsulation causes a marked decrease along the axes of iodine-125 seeds. In implants where the seed orientation is known (e.g., some high activity brain implants), this can be accounted for by using two-dimensional look-up tables. Usually the orientation is fairly random, as for example in a prostate implant containing 100 seeds. In this case the dose rate measured perpendicular to the seed is multiplied by an anisotropy factor, which gives an average dose rate for randomly oriented seeds.

T270. D Exposure rate = [(Exposure rate constant) $\times$ Activity]/(distance)^2

T271. A Because of the inverse square law, the PDD at a given depth from the surface of an applicator increases as the diameter increases.

T272. B Exposure rate at 100 cm = $8.25 \times 155 \times (1/100)^2$ R/hr = 127.9 mR/hr. 6 HVLs will reduce this by $(1/2)^6$ to 2.

B3.

T273. A Mean life $T_{avg} = 1.44 \times T_{1/2} = 86.4$ d = 2073.6 hr.
Total dose = (Initial dose rate) $\times$ (mean life)
18,000 cGy = (initial dose rate) $\times$ (2073.6 h) $\rightarrow$ Initial dose rate = 8.7 cGy/h

T274. A Total dose = initial dose rate $\times$ mean life, where mean life = half-life $\times$ 1.44
Total dose = $10 \times (60 \times 24) \times 1.44$ = 20736 cGy.

T275. A Total dose for a permanent implant
= initial dose rate $\times$ avg. life = initial dose rate $\times$ (1.44 $\times$ half-life)

B4.

T276.   B   The mean life of ^{198}Au = 2.7 days × 24 hr/day × 1.44 = 93.3 hr. Thus, if you assume the line is 50 cGy/hr, you would calculate the patient's total dose to be 50 × 93.3 = 46.7 Gy. If the line had been properly labeled as 50 Gy total, the patient would receive 7% more dose than you had assumed.

T277.   D   A Bq is the SI unit of activity and is equal to one disintegration per second.

T278.   C   1 Ci = 3.7 × 10^{10} Bq. 1 Bq = 2.7 × 10^{-11} Ci.

T279.   C   Although the gamma energies are different, the dose distributions in tissue around Ra and Cs sources are similar.

T280.   E   1 mg-Ra eq. of a radionuclide is that activity which gives the same exposure rate as a 1 mg source of Ra at the same distance. It is found by taking the ratio of the exposure rate constants.

T281.   A   Radium-226 has an exposure rate constant of 8.25 R-cm^2/mg-hr at 1 cm and cesium-137 has a constant of 3.26 R-cm^2/mCi-hr at 1 cm. The dose rate at 1 cm is the same for a 25 mCi source as for a 10 mg-Ra source.
8.25 × 10 = 3.26 × ?
? = 25 mCi

T282.   D   55 mg-Ra eq. = 55 × (8.25/3.26) = 139.2 mCi of cesium-137. A mg-Ra eq. of another isotope will have the same exposure rate as a mg of radium.

T283.   A   The exposure-rate constant for iridium-192 is 4.6 (R-cm^2/mCi-hr), and 8.25 (R-cm^2/mCi-hr) for radium. The dose rate will be higher by a factor of 8.25/4.6 = 1.79.

B5.

T284.   C   The magnifying ring should be placed in the same plane as the applicator for both films. The magnifying factor of the film is then (diam of ring on film)/(actual diam). Dimensions measured on the film are then divided by this magnifying factor to give actual dimensions in the plane of the applicator.

T285.   C   The distance (r) between two points in a cartesian coordinate system is:
$r = [(X_2 - X_1)^2 + (Y_2 - Y_1)^2 + (Z_2 - Z_1)^2]^{1/2}$
$= [(0 - (-2)^2 + (0 - 3)^2 + (2 - 4)^2]^{1/2}$
$= [(2)^2 + (-3)^2 + (2)^2]^{1/2} = (17)^{1/2} = 4.1$ cm.

T286.   D   Most treatment planning computers can use either shift films or a pair of orthogonal films to reconstruct the 3-D coordinates of the sources. Provided the films are orthogonal, they can be taken at any convenient angle.

B6.

T287.   C   Patterson-Parker rules require sources of equal activity/cm to be placed around the periphery of the square. A single central bar is required, so that the distance between lines is less than twice the treating distance of 1 cm. The activity/cm of the central bar is half that of the periphery. Thus four 4 mg needles and one 2 mg needle are used. Total activity = 18 mg, time = 3300/18 = 183.3 hrs.

T288.   E   Patterson-Parker rules specify sources of lower activity (mg/cm) at the center of an implant than those around the edge. If the rules are followed, the dose rate at a specified distance from the source plane will not vary by more than 10%. For equal activity sources and equal spacing, a system such as that devised by Quimby must be used.

T289.   A   In general, the Patterson-Parker rules require a non-uniform distribution of radium to produce a uniform dose distribution.

T290.   A   In order to maintain a uniform dose rate between adjacent planes, the number of milligram-hours must be increased as the spacing of the planes is increased.

T291.   B   Oblique filtration in radium tubes and needles is accounted for in the Patterson-Parker rules.

T292.   B   Although low dose rate irradiation may require an increased dose to obtain the same biological response as obtained with high dose rate irradiation, the RBE of radium gamma rays is generally taken to be unity.

T293.   A   A single plane implant requires 370 mg-hr to deliver 1000 cGy at 0.5 cm. Two planes separated by 1 cm deliver equal doses to the midplane, 0.5 cm from either plane.· Half of the 6000 cGy, 3000 cGy, will be delivered by each plane. Since the dose must be delivered in 6 d × 24 hr/d = 144 hr, the activity in each plane is equal to (3 × 370 mg-hr)/(144 hr) = 7.7 mg.

T294.   D   According to the Patterson-Parker rules an uncrossed end reduces the volume by 7.5%. If both ends are uncrossed, the volume is reduced by 15%.

T295.   B   The Patterson-Parker (Manchester) system uses a non-uniform distribution of sources to achieve a uniform dose at a given distance from the plane of the implant or throughout the volume.

T296.   C   The Paris system (Pierquin, Dutreix) is used for iridium-192 wires that are uniformly distributed in a parallel fashion. The treatment is prescribed to 85% of the average of the minimum doses between the sources in the central transverse plane.

T297.   A   The New York or Memorial Hospital system uses a nomograph to help plan the number of afterloading catheters to insert to achieve a certain dose rate.

T298.   C   The uniformity is stated to be ± 10%, if the rules are followed.

B7.

T299.   C   The long half-life of radium sources means that calculation of source decay is not necessary, and facilitates record keeping. With cesium sources insertion times become gradually longer, and eventually sources must be replaced.

T300.   C   Radium and cesium have very different half-lives (1600 yrs. vs. 30 yrs.) and cesium has a lower gamma energy (660 keV versus an average of about 1 MeV for Ra). However, since attenuation and scatter almost balance each other up to a few cm from the source, the inverse square law is the dominant factor in the shape of the distribution.

T301.   D   While B and C are true statements, they are irrelevant. Cesium-137 is less hazardous than radium-226 since radium-226 sources may leak radon gas.

B8.

T302.   B   5 mSv = 0.5 rem/yr. Assuming a 40 hour work week, this is equivalent to = 0.25 mrem/hr. The exposure rate at 10 m (= 1000 cm) is: Activity × Exp. rate const. for Ra × $(l/d)^2$ R/hr = 485 × 8.25 × $(1/1000)^2$ × 1000 mR/hr = 4 mR/hr. To reduce the exposure rate from 4 to 0.25 is a factor of 16, or 2^4. Therefore 4 HVLs are required.

T303.   B   A and C reduce the exposure by one half, but doubling the distance reduces the exposure to one quarter of its previous value, because of the inverse square law.

T304.   D   2 cm is 4 HVLs. This will reduce the dose rate by a factor of $2^4 = 16$.

B9.

T305.   C   Iodine-125 decays by electron capture to an excited state of ^{125}Te. From this it decays to the ground state by emission of 35.5 keV gammas. Most of these gammas are internally converted. Electron capture leaves a vacancy in the K shell, which gives rise to characteristic x-ray emission when the vacancy is filled. These x-rays have energies in the range 27-35 keV.

T306.   C   In beta decay, the available energy is divided between the beta particle and the neutrino, giving each a spectrum of energy.

T307.   A   Technically, the gamma transition is a metastable decay of barium-137m.

T308. D $\Gamma = C \ \Sigma \ n_i \times E_i$, where n is the fractional abundance and E is the gamma-ray energy. C is constant. The two gamma-rays of cobalt-60 are emitted in cascade during every cobalt-60 disintegration. n_1 and n_2 are both equal to 1. The one gamma-ray of cesium-137 is emitted in every disintegration of cesium-137 and also has an n of 1. The two gamma-rays of cobalt-60 are both approximately twice as energetic as the one gamma-ray of cesium-137. Therefore, Γ_{Co} is approximately four times Γ_{Cs}.

T309. E The process of radioactive decay of radium-226 to stable lead will liberate at least 49 gamma-rays from the daughter products with energies from 0.184 to 2.45 MeV. Radon, the first daughter product, is primarily an alpha emitter. The useful gamma-rays are from the daughter products further down in the chain in equilibrium with radium in a sealed container. Radium and radon are being phased out in radiation therapy for other long-lived radionuclides which are safer to handle.

T310. C

T311. D

T312. A

T313. B

T314. E

T315. C

T316. A

T317. E

T318. A Phosphorus-32 is a beta emitter; it is administered as an IV injection of about 3 mCi.

T319. C $t/T_{1/2} = 1 \text{ day}/60 \text{ days}$. $A/A_0 = \exp - (0.693t/T_{1/2}) = 0.99$

T320. B $t/T_{1/2} = 5 \text{ months}/30 \times 12 \text{ months}$.

T321. A $t/T_{1/2} = 1 \text{ month}/5.26 \times 12 \text{ months}$.

T322. C Cesium-137 is one of many radioactive isotopes produced when uranium reactor fuel rods fission. When a neutron is added to an isotope, the initial mass number is one less than the final, e.g., $^{59}Co + n = {}^{60}Co$. Cyclotron-produced radioisotopes are generally short lived, and often beta plus emitters. There are very few naturally occurring radioisotopes; some examples are radon and its daughters, which are part of the uranium decay chain.

P-32 ←

T323. A Radium-226 has a $T_{1/2}$ of 1600 years and decays <1% per year.

T324. C Cobalt-60 has a $T_{1/2}$ of 5.26 years and decays about 1% per month.

T325.	B	Cesium-137 has a $T_{1/2}$ of 30 years and decays 2.3% per year.
T326.	E	Iridium-192 has a $T_{1/2}$ of 74.2 days and decays 1% per day.
T327.	D	
T328.	B	
T329.	A	
T330.	C	
T331.	E	
T332.	D	Iodine-125 has a lower exposure rate constant, a shorter half-life (60.2 days compared to 74.2 days) and dosimetry has been more difficult than for iridium-192. The average photon energy is 0.028 MeV (^{192}Ir has an avg energy of 0.38 MeV) and gives less exposure to medical personnel. The RBE is higher for iodine-125.
T333.	B	
T334.	E	
T335.	A	
T336.	C	

B10.

T337.	E	Only the penetrating 2.27 MeV betas from yttrium are used to treat the eye. The low energy betas from strontium are absorbed in the foil.
T338.	D	^{90}Sr ($T_{1/2}$ = 28 y) decays to ^{90}Y ($T_{1/2}$ = 64 h) by low energy β-decay (0.54 MeV max). ^{90}Y in turn decays by higher energy emission (2.27 MeV max) and these βs are used for therapy, e.g., in ophthalmic applicators.
T339.	D	Strontium-90 decays to yttrium-90 by beta-minus decay, emitting a 0.54 MeV max β-ray which is mostly absorbed in the metal encapsulation of the applicator. Yttrium-90 subsequently undergoes a beta-minus decay of 2.27 MeV max. It is these beta-rays which are used in therapy.
T340.	E	^{90}Sr has a radioactive daughter, yttrium-90. It is the 2.27 MeV betas from ^{90}Y that are used therapeutically, not the 540 keV betas from ^{90}Sr, which do not penetrate very far.

B11.

T341.   B   Radium in a sealed container is an example of secular equilibrium since its long half-life means there is negligible decay in the time taken to establish equilibrium (about 1 month).

T342.   C   In secular equilibrium the daughter, which has a short half-life, appears to decay with the half-life of the parent. (This is because it is constantly being replaced by the parent as it decays.) Examples are radium decaying to radon and strontium-90 decaying to yttrium.

T343.   A

T344.   B

T345.   A

B12.

T346.   C   Due to the inverse square law, large diameter ovoids decrease the mucosal surface dose, but increase the depth dose. Since all ovoids (except mini ovoids) have the same internal shielding, doses to bladder and rectum are not affected.

B13.

T347.   B   A sensitive detector is the only immediate, foolproof way to verify that no sources have been left in the patient or dropped in the bed during removal. (A calibrated detector is not required since we are looking for the presence of activity and are not concerned with dose rate.) This should be done as soon as the sources have been removed because of the serious consequences to the patient if sources are left in longer than intended. The removed sources should be moved away from the patient's immediate area during the measurement.

T348.   C   Because the inverse square law causes a rapid fall-off of dose rate near a source, normal tissues around a brachytherapy implant can generally be spared to a greater extent than with an external beam which traverses normal tissue before reaching the tumor.

THERAPY Treatment Machines

TM1. Linear Accelerators

T349.   Which of the following does ***not*** apply to a flattening filter?
         A. must be very carefully centered on the beam axis
         B. is needed to correct for the forward peaked dose distribution created by electrons striking a
            target
         C. is always used in megavoltage x-ray beams
         D. is always used in megavoltage electron beams → *use scattering foils*
         E. can be subject to activation, depending on x-ray energy

T350.   With regard to the production of electron beams by linear accelerators, which of the following
         is true?
         A. the beam current is much higher in the "electron mode" compared with the "photon mode"
         B. electron beam flatness depends on the design of the cone or applicator
         C. the bending magnet is rotated out of the beam when "electrons" are selected *NO*
         D. thick scattering foils can be used to reduce bremsstrahlung
         E. all of the above

T351.   Linear accelerators require flattening filters to:
         A. filter out low energy photons *← yes but not purpose*
         B. reduce dose rates to a safe level
         C. decrease the intensity at the center of the beam
         D. compensate for the pulsing of the radiation
         E. increase the penetrating ability of the beam *← yes but not purpose*

T352.   The reason an accelerator contains two ionization chambers is to:
         A. measure energy degradation of the beam
         B. quantify the velocity of the photons or electrons
         C. for back up in case of failure
         D. one measures output and the other measures flatness and symmetry
         E. all of the above

T353.   The purpose of the scattering foil in the electron mode of a linear accelerator is to:
         A. absorb scattered electrons
         B. change the x-ray beam into electrons
         C. shield the ion chambers
         D. absorb excess radiofrequency energy
         E. provide homogeneity across the electron field

T354. A 20 MeV electron beam strikes a standard transmission target in a linear accelerator. The fraction of the total energy of the beam that will be converted into x-rays will decrease if the:
A. electron beam energy (E) is increased
B. atomic number (Z) of the target material is increased
C. target is made thinner
D. diameter of the target is doubled
E. target angle is decreased

T355-360. Match the four sections of a linear accelerator with their components or functions (answers may be used more than once).
A. modulator
B. microwave power
C. accelerator
D. beam handling

T355. C Side coupling cavity

T356. B Magnetron

T357. D Bending magnet

T358. A Thyratron

T359. D Monitor chamber

T360. B Beam energy control

T361. Bending magnets in linear accelerators:
A. are not required when the accelerator waveguide is parallel to the floor
B. are designed to bend the photon beam 90° or 270°
C. are designed to bend the electron beam 90° or 270°
D. are designed to bend the photon beam 180°
E. none of the above

T362. For a particular treatment protocol the dose rate on a linac must be reduced; this can be done by changing the:
A. AFC (automatic frequency control)
B. microwave frequency
C. machine calibration (cGy/MU)
D. voltage across the monitor ion chamber
E. PRF (pulse repetition frequency)

T363. The advantage of a beam stopper on a treatment machine is:
A. it makes coronal fields easier to set up
B. it reduces the dose at the operator's console
C. it reduces the wall thickness in the direction of the primary beam only
D. it reduces the use factor for lateral beams
E. it reduces the thickness of all the walls in the treatment room

T364. Which of the following does **not** occur when a linac is changed from the x-ray mode to the electron mode? (excluding units with scanned electron beams)
A. the target is removed
B. a scattering foil is placed in the beam
C. the monitor chamber is removed
D. an electron applicator is attached
E. the beam current decreases

T365. When electrons are accelerated in a linac waveguide from 10 MeV to 20 MeV:
A. their velocity increases
B. their potential energy increases
C. they can travel faster than the speed of light
D. their mass increases
E. their mass decreases and their kinetic energy increases

T366. A monitor unit (MU) is:
A. a length of time (measured in minutes) for which a linac beam is turned on
B. affected by changes in dose rate
C. related to the dose delivered at a reference point
D. a unit used to describe the electron energy of a linear accelerator
E. part of the hardware of a record and verify system

T367. Klystrons and magnetrons are:
A. devices used to bend beams of electrons
B. located in the treatment head of a linac
C. beam focusing devices
D. sources of microwave power
E. part of the timer circuit of a linear accelerator

T368. In therapy accelerators the monitor chambers are used to:
A. terminate treatment after set dose is delivered
B. provide feedback about beam symmetry
C. control the beam's energy
D. B and C only
E. A and B only

T369. Components of a linear accelerator treatment head that are used in the x-ray mode only and not in the electron mode are:
A. scattering foils and target
B. target and flattening filter
C. primary collimators and monitor chamber
D. monitor chamber and cones
E. magnetron and field light

T370. A flattening filter:
A. compensates for patient irregularities
B. is designed to decrease depth dose
C. is of uniform thickness across the width of the beam
D. is used in the photon mode only
E. is used to broaden the electron field

T371.   Which of the following does *not* accelerate electrons?
A. 6 MV medical linac
B. cyclotron — *accelerates (+) charged particles*
C. betatron
D. microtron
E. orthovoltage unit

T372.   To produce a photon beam from a linear accelerator, in what order are the following components arranged in the head?
1. flattening filter
2. scattering foil
3. target   *3, 1, 4*
4. monitor chamber

A. 3, 2, 4
B. 4, 3, 1
C. 3, 2, 1, 4
D. 3, 1, 4
E. 1, 3, 4

T373.   For a 6 MV linear accelerator, the:
A. 6 MV is applied between the filament and the target
B. average photon energy within the beam is 6 MeV   *most will be < 6 meV*
C. photons all have an energy of 6 MeV
D. photon beam is twice as penetrating as a cobalt-60 beam (3 MeV equivalent)
E. maximum energy acquired by an electron in the wave guide is 6 MeV

T374.   When 20 MeV electrons are fired at a target the x-ray production is most intense at ___ degrees to the direction of the electron beam.
A. 0
B. 30
C. 45
D. 90
E. it is equal in all directions

TM2. Cobalt-60 Teletherapy Units
open & closed position

T375.   Regarding the travel time of a cobalt-60 source of a modern teletherapy machine, all of the following are true *except:*
A. should be considered in the daily administered dose to the patient
B. requires a timer correction in the order of ± 0.02 sec   *min*   *1-2 sec*
C. is unimportant because of the short treatment times
D. differs from machine to machine
E. affects the first and last seconds of treatment

T376.   The difference in "timer error" between two units could be due to a difference in:
A. source diameters
B. source travel time
C. isocenter distance
D. dose rate

T377.   Machine output monitor chambers are *not* used in cobalt-60 units because:
A. they are only used for electrons
B. cobalt-60 has a predictable output
C. temperature-pressure corrections cannot be made on a cobalt-60 unit
D. there is no room in the head of the machine
E. they will perturb the cobalt-60 beam

TM3. Superficial X-Ray Units

T378.   If a 2 mm Al filter is replaced by a 1 mm Al filter on a superficial x-ray machine the effect will be to:
A. harden the beam
B. reduce the dose rate at d_{max}
C. increase the PDD at 1 cm depth
D. all of the above
E. none of the above

T379.   It is found that a superficial treatment was performed with a 2 mm Al filter instead of a 3 mm Al filter (using the time calculated for the 3 mm filter). The effect on the patient would be:
A. increased skin dose but no change in PDD
B. increased skin dose and increased PDD
C. increased skin dose and decreased PDD
D. decreased skin dose and decreased PDD
E. decreased skin dose and increased PDD

T380-381.   Match the voltage ranges with the appropriate materials in which the HVL (half-value layer) is measured:
A. mm Pb
B. mm Cu
C. mm Al
D. mm Lucite

T380.   10 kV – 120 kV

T381.   120 kV – 1 MV

TM4. NRC Regulations
(part 35)

last update

→ regulate radioactive material HDR/LDR coCo
not acceleration

Agreement states
State regulates
for radioactive
material
→ enforce
regulations
that are
the same
as NRC

T382. When a ^{60}Co source is replaced with one of higher activity but of the same diameter, all of the following must be done prior to its continued use *except*:

A. the output must be measured with a calibrated ionization chamber, and must not differ by more than 5% from that stated by the manufacturer

B. a complete set of beam data (CAX depth dose, profiles) must be reacquired using a scanning probe in a water tank

C. the head must be surveyed to ensure that the dose rate at 1 m from the source does not exceed 10 mrem/hr max, and 2 mrem/hr average

D. a radiation survey must be performed outside the treatment room, in all occupied or potentially occupied areas

E. light vs. radiation symmetry and alignment must be verified

State –
regulators for
radioactive
materials and
accelerators

T383. According to NRC regulations, a cobalt-60 unit used for patient treatment must have all of the following *except*:

A. the unit must be equipped with a door interlock

B. the technologist must be in visual contact with the patient at all times

C. the unit must be equipped with a record and verify system

D. the attending physician must have a license for application of radioactive materials

E. the output must have been calibrated within the last year

T384. According to NRC Part 35, if the therapy dose delivered differs by more than 10% from the final prescription, all of the following must be reported to the NRC (Nuclear Regulatory Commission) *except* radiation dose from:

A. a brachytherapy source

B. a cobalt-60 source

C. a linear accelerator

D. a radiopharmaceutical therapy dose

E. none of the above since the regulation is for a 50% difference from the prescription

T385. According to the NRC the following should be checked monthly as part of a QA program:

1. light/radiation field coincidence
2. output factors for the range of field sizes in use
3. timer error and linearity
4. wedge transmission

A. 1, 2, 3

B. 1, 3

C. 2, 4

D. 4 only

E. all of the above

NCRP/ICRP
BEIR/UNSCEAR → make recommendations
collect data

NRC - make regulations +
enforce

FDA - regulates manmade
equipment - accelerators
x-ray equipment
manufacturing
not calibration

T386. According to 1992 NRC regulations, the output for one set of operating conditions for a cobalt-60 unit must be measured:
A. daily
B. weekly
C. monthly
D. yearly
E. every 5 years

T387. According to the NRC, monthly safety checks include all of the following *except*:
A. electrical interlocks at the room entrance
B. beam on-off indicator lights
C. viewing systems
D. accuracy of distance measuring devices
E. maximum and average exposures at 1 meter from the source

T388. The NRC requires that all survey meters in use be calibrated:
A. annually, at one point
B. every 6 months at one point on each scale
C. annually at 2 points on each scale
D. every 2 years at 2 points on each scale
E. annually at one point on each scale

T389. All of the following are true about a cobalt-60 teletherapy source change *except*:
A. the unit must be fully inspected and serviced by a specially licensed person
B. the source must be leak tested
C. radiation surveys, such as maximum and average dose rates at 1 meter form the source, must be carried out
D. full calibration and safety checks must be performed
E. additional room shielding will usually be necessary

T390. Amendments to a teletherapy license must be made for which of the following?
A. when changes are made in room shielding
B. when the teletherapy unit is relocated
C. when the unit is moved within the room
D. when the physicist on the license is changed
E. all of the above

T391. According to NRC regulations Part 35 all of the following safety features of a cobalt-60 teletherapy unit are necessary *except*:
A. the room door must be closed in order to turn the beam on
B. when the room door is opened, the beam is turned off automatically
C. there is a beam condition indicator light outside the room
D. there must be a radiation monitor in the room with a backup battery
E. the source position indicator rod must be visible on the TV monitor

T392.   The NRC's definition of "full calibration" of a cobalt-60 teletherapy unit requires all of the following *except:*
         A. the calibration must be supervised by, but need not be done by, the teletherapy physicist named on the license
         B. measurement of output for the range of field sizes in use
         C. light-radiation coincidence
         D. field uniformity at several gantry angles
         E. on-off error

T393-396. Match the frequency with the procedure, according to the latest NRC regulations:
         A. daily
         B. weekly
         C. monthly
         D. quarterly
         E. annually

T393. E   Full calibration of cobalt-60 teletherapy unit. E

T394. C   Safety spot checks on cobalt-60 teletherapy unit. C

T395. D   Brachytherapy sealed source inventory. B

T396. A   Verification of operation of teletherapy room monitor. A

T397.   NRC regulations apply to the use of:
         1. Cobalt teletherapy units
         2. Brachytherapy sources
         3. Linear accelerators
         4. Diagnostic x-ray machines

         A. 1 only
         B. 1, 3
         C. 1, 2, 3, 4
         D. 1, 2
         E. 1, 3, 4

TM5. Neutron Production in Linacs

T398.   Regarding neutrons, which of the following is *false*?
         A. neutrons are produced in therapy linacs operating at nominal energies of 10 MeV and above
         B. additional shielding for neutrons is required for linacs operating at nominal energies of 15 MeV and above
         C. therapeutic neutron beams have a depth dose curve which most closely resembles that of a proton beam
         D. neutrons have a high "quality factor" because they give rise to "knock-on" protons
         E. neutrons are more readily attenuated by concrete than by an equal mass of lead

T399.   The photon beam energy at which neutron contamination of the beam and neutron dose to
        personnel must be considered is ___ MeV.
        A. below 2
        B. 4
        C. 6
        D. 8
        E. above 10

        need neutron shielding ≳ 15mv

T400.   In practice, the minimum accelerator energy for which additional neutron shielding is required
        in a treatment room to reduce the dose from (γ, n) reactions is ___ MV.
        A. 4
        B. 10
        C. 15
        D. 25
        E. 45

T401.   Neutrons produced as leakage by therapy linacs:
        A. require extra room shielding above a nominal photon energy of 8 MeV
        B. are slowed down by materials containing hydrogen
        C. are most effectively shielded with lead
        D. can be measured with the same exposure meter used to measure photon leakage
        E. all of the above

T402.   Polyethylene is sometimes found on entrance doors to high energy accelerators in order to:
        A. absorb low energy photons
        B. make the door lighter
        C. absorb high energy neutrons
        D. scatter and thermalize high energy neutrons

        Boron captures neutrons once thermalized

T403.   Most of the neutrons produced by a high energy linac are produced:
        A. at the collimator and target area   *high Z materials*
        B. in the patient
        C. in the floor of the room
        D. in the electron mode rather than the photon mode

TM6. Other Treatment Modalities

T404.   The potential advantage of treating with a 14 MeV neutron beam is:
        A. higher RBE and lower OER
        B. lower RBE and higher OER
        C. higher RBE and OER
        D. lower RBE and OER
        E. a nearly ideal depth dose distribution

T405. When 6 MeV neutrons interact with tissue:
A. they have a range (in cm) equal to about 1/2 their energy (in MeV)
B. the decrease of dose rate with depth is exponential and is similar to that of 250 keV x-rays
C. 6 MeV neutrons are not easily attenuated and have a depth dose curve similar to that of 20 MV x-rays
D. 6 MeV neutrons have a similar depth dose curve to 20 MeV electrons
E. the depth dose curve most closely resembles that of a 100 MeV proton beam

T406. Which of the following properties is *not* associated with heavy particle beams?
A. high LET compared to electrons
B. Bragg peak
C. energy dependent stopping power
D. fixed range in tissue for fixed initial energy
E. uniform depth dose over its entire range

T407. Therapeutic proton beams operate in the ___ MeV range:
A. 0.25
B. 2.5
C. 25
D. 250
E. 2500

T408. Fast neutrons are being considered as a useful treatment modality because of their:
A. much better depth dose distribution than high energy x-rays
B. lower dependence on cell cycle and minimization of excessive repair capacity
C. greater dependence on oxygen levels
D. loss of the major portion of their energy at the end of their range
E. ability to adjust range in tissue by altering energy

TM7. Quality Assurance

T409. Typical beam flatness specifications for a linac photon beam are:
A. $\pm$ 10% over the whole width of the beam measured in air at the isocenter
B. $\pm$ 6% over 50% of the field width measured in water at d_{max}
C. $\pm$ 3% over 80% of the field width measured in water at 10 cm depth
D. $\pm$1% over 90% of the field width measured in air at the isocenter
E. $\pm$ 20% over 50% of the field width measured at the isocenter in air

T410. During QA tests on linacs, beam symmetry and flatness can be checked using several different methods. However, the baseline data measured during acceptance testing of the new linac must be measured with:
A. film exposed parallel to the beam axis and evaluated using a scanning densitometer
B. an array of diodes providing a real-time display of the dose profile
C. a small volume ion chamber scanned across a water tank
D. a calibrated thimble chamber with a build-up cap exposed in air at 1 cm intervals across the beam at the level of the isocenter

T411. According to NCRP Report 69, the alignment of the light beam and the therapy beam on any side of a 10 × 10 cm field **shall not** exceed:
A. 1 mm
B. 2 mm
C. 3 mm
D. 4 mm
E. 5 mm

TM8. Calibration

T412. The protocol currently recommended by the AAPM for calibrating megavoltage therapy units is:
A. SCRAD
B. the Nordic protocol
C. TG-21 · task grap 21
D. ICRU Report 21

T413. During the calibration of a cobalt-60 unit, the temperature-pressure correction is omitted. If P = 776 mmHg, and t = 22 °C, the stated output will be about: (note: STP is P = 760 mmHg, t = 22 °C). same
A. 1% low
B. 2% low
C. 4% low
D. 2% high
E. 3.5% high

$$\left(\frac{760}{776}\right)\left(\frac{273+22}{295}\right) = .979$$

$$\left(\frac{760}{P}\right)\left(\frac{273+t}{295}\right) = correction \sim 2\% low$$

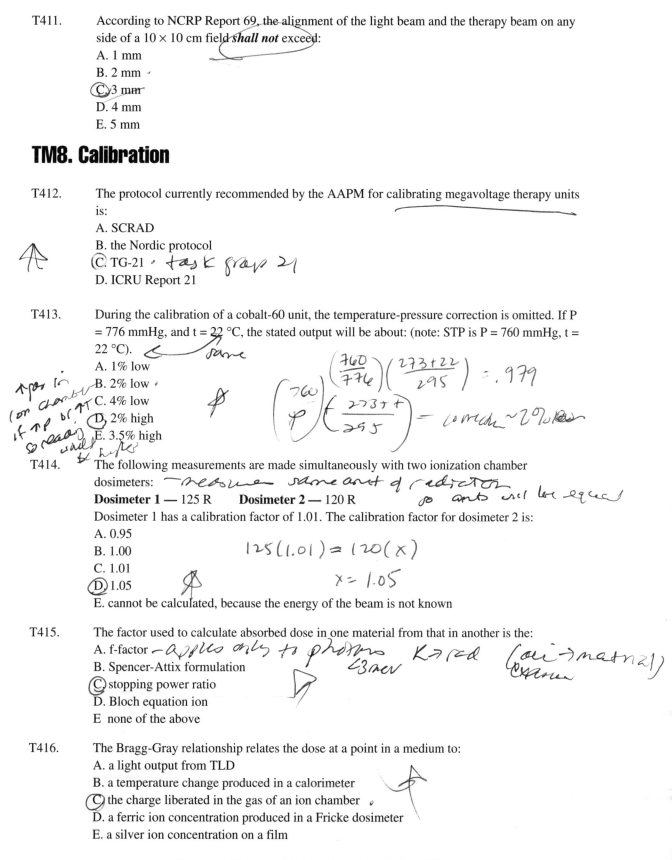

T414. The following measurements are made simultaneously with two ionization chamber dosimeters: measure same amt of radiation so amts will be equal
Dosimeter 1 — 125 R   **Dosimeter 2** — 120 R
Dosimeter 1 has a calibration factor of 1.01. The calibration factor for dosimeter 2 is:
A. 0.95
B. 1.00
C. 1.01
D. 1.05
E. cannot be calculated, because the energy of the beam is not known

$$125(1.01) = 120(x)$$
$$x = 1.05$$

T415. The factor used to calculate absorbed dose in one material from that in another is the:
A. f-factor — applies only to photons K→rad (air→material) <3MeV exposure
B. Spencer-Attix formulation
C. stopping power ratio
D. Bloch equation ion
E none of the above

T416. The Bragg-Gray relationship relates the dose at a point in a medium to:
A. a light output from TLD
B. a temperature change produced in a calorimeter
C. the charge liberated in the gas of an ion chamber
D. a ferric ion concentration produced in a Fricke dosimeter
E. a silver ion concentration on a film

T417.   A physicist measures the output of a linac and finds it to be 4% low. The usual action taken by the physicist is:

A. to change the tables of output factors (cGy/MU for all collimator settings) to the new measured values

B. if the calibration is within ± 4%, nothing is changed

C. all patients treated within the last month must be notified

D. a potentiometer is adjusted so that one monitor chamber unit is equal to one cGy measured at the reference point

E. none of the above

T418.   In order to calibrate a 10 MV x-ray beam, the ionization chamber used must have a calibration factor. This factor:

1. is measured in a cobalt-60 beam

2. must be measured using the same energy for which it will be used, in this case 10 MV x-rays

3. must have been measured at an accredited dosimetry calibration laboratory (ADCL) within the last year

4. must have been measured at an ADCL within the last two years

A. 1, 3

B. 1, 4

C. 2, 3

D. 2, 4

E. none of the above

T419.   When calibrating a cobalt-60 therapy unit, the temperature and pressure are recorded. This is because:

A. the temperature and pressure affect the output of the ^{60}Co source

B. the calibration equipment can only be used within a range of about 5°C and 10 mm mercury, depending on the manufacturer

C. the mass of air in the ionization chamber will be greater at high temperature and high pressure

D. the mass of air in the ionization chamber will be greater at low temperature and low pressure

E. none of the above

T420.   Bragg-Gray theory is used in the derivation of:

A. the theoretical shape of the % depth dose curve for high energy proton beams

B. the f-factor

C. the number of mg-Ra eq. required per 1000 cGy in a single plane implant

D. the thickness of the primary barrier in a cobalt-60 room

E. factors used in photon dosimetry protocols for converting ion chamber readings to dose in a medium

T421.   Which of the following is a thimble-type chamber?
         A. free-air ionization chamber
         B. extrapolation chamber
         C. Farmer chamber
         D. silicon diode
         E. parallel-plate chamber

T422.   An unsealed ion chamber will have a lower charge reading in city A than an unsealed chamber
         exposed to the same amount of radiation in city B because:
         A. the relative humidity might be higher in city B
         B. the temperature might be higher in city B
         C. the pressure might be lower in city A
         D. the stem effect might be more in city B
         E. background radiation could be higher in city B

T423.   A build-up cap for an ionization chamber:
         A. is used to protect the chamber from mechanical damage
         B. should be made of a high-Z material
         C. prevents stem leakage
         D. corrects for perturbations in the beam
         E. must be thick enough to provide, along with the chamber wall, electronic equilibrium

T424.   An air equivalent chamber means:
         A. the effective Z of the chamber wall is similar to that of air
         B. the Z of the wall is approximately 7.6
         C. the number of electrons/gm is the same as that of "solid" air
         D. all of the above

T425.   An instrument used to measure "absolute exposure" (i.e., need not be calibrated against a
         standard chamber) is the:
         A. Baldwin-Farmer ionization chamber
         B. Victoreen condenser chamber
         C. free-air chamber
         D. TLD
         E. diode

T426.   Which of the following corrections are commonly made for dose determinations with ion
         chambers?
         1. temperature and pressure correction
         2. correction for loss of ionization due to recombination
         3. correction for chamber volume
         4. correction for W value

         A. 1 and 2
         B. 1 and 3
         C. 1, 2, 3
         D. all of the above

T427. The nominal energy of a megavoltage photon beam (required to evaluate certain parameters used in the TG-21 calibration protocol) is defined by:
A. the half-value layer in lead — *dark case HVL in megavoltage*
B. the energy stated by the manufacturer
C. the stopping power ratio
D. the maximum energy of the electrons at the end of the waveguide
E. the ratio of TMRs at 10 and 20 cm depths

T428. An ionization chamber is exposed to 100 R at 22°C in New York. The same chamber is then exposed to 100 R in an identical beam at 22°C in Denver. The reading will be:
A. higher in New York
B. higher in Denver
C. the same in both locations

↑ altitude ↓ pressure
so Denver will have a higher C_{TP}

TM9. Head Leakage

T429. The leakage at 1 meter lateral to the treatment head of a 4 MeV accelerator **should** be less than _____ of the useful beam at the isocenter.
A. 0.001%
B. 0.01%
C. 0.1%
D. 1%
E. 10%

T430-433. Match the following conditions with the dose rate:
A. 1 R/hr at 1 meter
B. 5 mR in any one hour
C. 0.1% primary beam intensity at 1 meter
D. average 2 mR/hr at 1 meter
E. 100 mR in any one hour

2mR/h allowed @ treatment console

T430. D Leakage for cobalt-60 in the "off" position

T431. C Leakage for cobalt-60 in the "on" position

T432. B Radiation area requiring posting

T433. E High radiation area requiring posting

also for linacs *memorize these*

T434. A pregnant technologist is concerned about fetal exposure when working on the cobalt-60 unit. Assuming (as a worst case scenario) the maximum legal head leakage at 1 meter from the source, how much time per week could the technologist remain at this location without exceeding the MPD to the fetus?
A. 1 min
B. 10 min
C. 1 hr
D. 10 hr
E. 40 hr

OCC

50mSv = 5 rem
lens 150 = 15
OTU 500 = 500

Answers

TM1.

T349.   D   Flattening filters are only used in photon beams; electrons use scattering foils to obtain a large field size.

T350.   B   The beam current is higher in the photon mode. The bending magnet is required to point the electrons towards the isocenter; it is positioned prior to the x-ray target or the electron scattering foil. The thicker the scattering foil, the greater the number of interactions and the higher the % of bremsstrahlung. Cone design and collimator offset are both important for electron beam flatness.   *impt*

T351.   C   The maximum dose rates of a linear accelerator are found along the central axis with rapidly decreasing dose rates laterally. Beam flattening filters are used to compensate for this effect.

T352.   C   Separate chambers with separate circuits are used in the accelerator to assure reliable monitoring of radiation output.

T353.   E   The scattering foil changes the narrow electron beam into a broader beam by forcing the electron paths to change due to collision in the foil.

T354.   C   A standard transmission target is thick enough to absorb a large fraction of the electron beam and thin enough to not appreciably lower the x-ray output. A thinner target would absorb fewer electrons and produce fewer x-rays. The fraction of the total energy of an electron beam that will be converted into x-rays is proportional to $Z \times E$, where Z is the atomic number of the target and E is the kinetic energy of the electrons. Target angle applies to kilovoltage x-ray machines.   *impt*

T355.   C   Accelerator waveguides designed for standing waves have alternate cavities of the waveguide displaced to the sides. This reduces the total length of the waveguide and may obviate the need for a bending magnet.

T356.   B   The magnetron is a microwave generator. The microwaves are fed to the accelerator waveguide via waveguides filled with pressurized freon gas. In high energy accelerators, the magnetron may be replaced by a klystron.

T357.   D   If the waveguide is too long to be mounted isocentrically in a straight through configuration, it will be mounted parallel to the gantry rotation axis, and the electron beam will be turned 90° or 270° by a bending magnet to point at the isocenter.

T358.	A	The modulator includes a pulse forming network (PFN) and a switching tube. This is known as a thyratron. Typically the electron gun and magnetron are "switched on" about 200 times per second, delivering a pulsed beam of photons.
T359.	D	The chamber is positioned in the head, after the target and the flattening filter. It monitors dose delivered, and beam symmetry and flatness.
T360.	B	Beam energy is affected by the frequency of the microwaves filling the accelerator tube. The automatic frequency control (AFC) maintains this frequency at the optimal level.
T361.	C	The electron beam leaves the evacuated accelerator waveguide and is turned through 90° or 270° before striking the target for the x-ray mode. (X-rays cannot be bent by magnetic fields).
T362.	E	The dose rate is proportional to the pulse repetition frequency (PRF).
T363.	C	A beam stopper reduces the primary beam by a factor of 1000. The other walls not in the primary beam, which shield against scatter and leakage, are unaffected.
T364.	C	The monitor chamber remains in position to monitor the beam output but the target and flattening filter are replaced with a scattering foil. The electron applicator is usually interlocked so that the electron beam cannot be turned on unless it is attached.
T365.	D	Electrons at 10 MeV are already traveling at almost the speed of light, so an increase in energy results in an increase in mass.
T366.	C	1 monitor unit is the amount of charge collected by the monitor ionization chamber through which the beam passes in a linac. It represents an amount of dose delivered to a reference point (e.g., 1 cGy per MU for a 10×10 cm field at 100 cm SSD, d_{max}). It is independent of dose rate. The time taken to deliver a dose, however, will depend on the dose rate. On linacs, the timer setting is used only as a back-up in the unlikely event that both monitor chambers fail.
T367.	D	Magnetrons generate microwaves and are usually used in the lower energy range. Klystrons are microwave amplifiers; they are more expensive than magnetrons but last longer and are generally used in the higher energy range.
T368.	E	Sections of the monitor chamber at equal distances from the center are used to monitor symmetry.
T369.	B	In x-ray mode, the target is struck by the electron beam to produce x-rays with beam intensity peaked forward. The intensity is made uniform by a flattening filter. Scattering foils are used in electron mode to spread out the intensity of the electron beam. Primary collimators and monitor chambers are used in both modes as are the magnetron (produces microwaves) and the field light (used for localization).

T370.　D　A high energy photon beam is peaked forward and must be flattened. The flattening filter is shaped like an ice cream cone and differentially attenuates across the beam diameter.

T371.　B　The linac accelerates electrons in a linear accelerating waveguide. The betatron accelerates electrons (in a circular orbit) with a changing magnetic field. A microtron is an electron accelerator which combines the principles of the linac and the betatron. An orthovoltage unit has an x-ray tube in which electrons are emitted from the filament and accelerated toward the anode when high voltage is applied between the anode and the cathode. The cyclotron accelerates positively charged particles.

T372.　D　The scattering foil replaces the flattening filter in the electron mode. The flattening filter flattens the beam coming out of the target, and the monitor chamber checks this beam for flatness and symmetry and turns off the beam after the correct number of MU have been delivered.

T373.　E　The waveguide uses microwaves to accelerate electrons to 6 MeV. The target of a linear accelerator is not connected to a high voltage as in a conventional x-ray tube. Most of the photons will have energies less than 6 MeV.

T374.　A　Although x-ray intensity is greatest in a lateral direction for low energy x-rays (hence the use of a "reflection" target), as energy increases x-ray production tends towards the forward direction; thus linacs use transmission targets.

TM2.

T375.　C　A cobalt-60 source must be shielded when it is not in use. During treatment the source must be unshielded. This is done by either moving the source from a shielded position to an open position or by removing the shield from in front of the source. In either case a finite time, in the order of a few seconds, is required to unshield the source at the beginning of treatment and to shield the source at the end of treatment. Since for modern cobalt-60 teletherapy units, treatment times are in the order of a minute or less, the dose delivered during source transit must be taken into account.

T376.　B　Timer error, usually ± 0.01 to 0.02 minutes, is affected by the source travel time and the timer mechanism.

T377.　B　An ion chamber in the head of a linear accelerator monitors and checks for perturbations in dose rate, dose and field symmetry. The ^{60}Co unit has a steady and predictable output and does not need to be monitored constantly. However, the ^{60}Co output is checked monthly as part of the QA program.

TM3.

T378.   E   A, B, and C would occur if the filter thickness were increased.

T379.   C   A reduction in the filter thickness will increase the machine output and hence the skin dose. However, a less filtered beam has a lower HVL (relatively more low energy photons which have not been filtered out), so the PDD (as a percentage of the skin dose) will be less. The patient may, however, receive more total dose at a particular depth due to the increase in output.

T380.   C

T381.   B

TM4.

T382.   B   Re-acquiring a complete set of beam data is not necessary if the source diameter is the same size. The checks listed must be done to ensure that the source is positioned correctly and that the leakage is still within acceptable limits.

T383.   C   Although desirable, a record and verify system is not a requirement.

T384.   C   The NRC regulates the medical use of by-product materials which includes sealed sources (brachytherapy and cobalt-60 sources) but does not include linear accelerators. Problems with linear accelerators are reported to (not regulated by) the U.S. Pharmacopoeia but overdoses given with linear accelerators are not currently (1994) reported to either body.

T385.   B   2 and 4 are checked annually.

T386.   C   A "full calibration" is required annually but a monthly spot check of output is also required.

T387.   E   The exposure rates at 1 meter from the source (with the source off) are measured initially before clinical use, and after any source change or other change requiring a license amendment. The limits are 10 mrem/hr max and 2 mrem/hr avg at 1 meter.

T388.   C   The calibration date and exposure rate from a dedicated check source must also be conspicuously noted on the instrument.

T389.   E   Answers A through D are required by NRC Part 35 for a source change and a report is sent within 30 days to the licensing agency. Modification of shielding is almost never needed; the old, decayed source is replaced with one with a similar acuity to the previous source when new, for which the shielding was designed.

T390. E Changes in the shielding and location of the teletherapy unit can result in unacceptable dose rate levels outside the room. These changes must be approved. When the physicist or the radiation safety officer on the license is changed, the qualifications of the new person must be checked and approval given by the licensing agency.

T391. E The beam condition light and the radiation monitor indicate whether the beam is on and, although useful in an emergency, it is not a requirement to be able to see the source rod from outside the room.

T392. A The full calibration *must* be performed by the teletherapy physicist. Monthly spot checks can be performed by others under the physicist's supervision.

T393. E

T394. C

T395. D

T396. A See NRC's Code of Federal Regulations 10 CFR Part 35.

T397. D The NRC regulates only the human use of reactor by-products.

TM5.

T398. C Therapy neutron beams have a PDD between that of cobalt-60 and 250 kV x-rays, limiting their use for treating deep seated lesions.

T399. E The usual threshold cited for gamma-neutron interaction is about 8 MeV, but in practice neutrons are measurable at nominal photon beam energies of 10 MeV and above.

T400. C (γ, n) reactions have a threshold at about 8 MeV. However, the neutron dose is minimal at this energy and any neutrons produced are absorbed in the existing shielding. At about 15 MeV, extra shielding becomes necessary in the maze walls and door to reduce the neutron dose.

T401. B Neutrons are most effectively slowed down by striking (and sharing their energy with) particles of a similar mass, i.e., protons. The hydrogen nucleus is a proton, so hydrogenous materials are effective neutron moderators.

T402. D Polyethylene will scatter high energy neutrons that are produced by high energy accelerators (> 10 MV). The scattering reduces the neutron energy to thermal levels and reduces the dose due to neutrons. The polyethylene can also absorb thermal neutrons releasing gamma rays (through neutron capture).

T403.   A   Neutrons are produced by the interaction of high energy photons with high Z materials in the head of the machine.

TM6.

T404.   A   The depth dose is similar to ^{60}Co: neutrons undergo exponential attenuation. Charged particle beams such as protons have a more nearly ideal depth dose.

T405.   B   Neutrons are subject to exponential attenuation; "range" is associated only with charged particles.

T406.   E   Heavy particles have higher linear energy transfer [LET(keV/μm)] than electrons and have an increased rate of energy loss at the end of their range (the Bragg peak). Stopping powers and ranges are dependent on the energy.

T407.   D   The 250 MeV range allows penetration to about 38 cm in water.

T408.   B   D and E apply to protons. The depth dose is similar to a cobalt-60 beam and holds no advantage over x-rays. However, neutrons have less dependence on cell cycle effects than photons and have a less pronounced shoulder on the cell survival curve.

TM7.

T409.   C

T410.   C   The baseline data must be taken in a water tank providing full scatter. Any initial adjustments are made based on these scans, and a final set are taken and kept for future comparison. Baseline data can then be taken with the QA equipment which will be used routinely for spot checks. This could be film, or various arrays of ion chambers or diodes.

T411.   C   NCRP Report 69 states that the light/radiation coincidence *should* be better than 2 mm but *shall not* exceed 3 mm.

TM8.

T412.   C   This protocol, produced by the AAPM (American Association of Physicists in Medicine) Task Group 21, was published in *Medical Physics*, the journal of the AAPM, in Dec. 1983. It includes appropriate correction factors to account for the different types of chambers used by physicists.

ANSWERS

T413. D The temperature-pressure correction applied to the chamber reading is:
$(760 /P) \times [(273.15 + t)/295.15]$
In this case there is no temperature correction, as 22 °C is the standard temperature. 760/776 = 0.98, i.e., the corrected reading should be 2% lower, so the uncorrected output will be 2% high. (t is in °C and P is in mm mercury.)

T414. D Ionization chamber dosimeter calibrations are traceable to the National Bureau of Standards (NBS). The true exposure is the dosimeter reading corrected for temperature and pressure multiplied by the calibration factor. The calibration factor for dosimeter 2 is equal to $1.01 \times (125/120) = 1.05$. The dosimeter scale reading is smaller than the actual exposure.

T415. C Stopping power ratio relates interaction of radiation in one medium vs. another.

T416. C The Bragg-Gray relationship describes gas cavity interactions with x-rays.

T417. D If the dose delivered on a cobalt-60 unit differs from the prescribed dose by 10%, the patient must be notified. It is good practice to measure the output monthly and adjust the calibration of the monitor chamber if it has drifted by more than 2%.

T418. B Ionization chambers must be calibrated not less than every two years at an accredited dosimetry calibration laboratory. The calculation of cGy/MU for beams above ^{60}Co energy is based on Bragg-Gray cavity theory, and involves the calibration factor for ^{60}Co.

$PV = nRT$

T419. E The mass of air in the ionization chamber will be greater at low temperature and high pressure, giving a greater reading (more ionization) than at 22 °C and 760 mm mercury. The reading is therefore corrected back to the value that would have been obtained at the nominal temperature and pressure, at which the calibration factor is correct.

b/c:
$\left(\dfrac{760}{p}\right)\left(\dfrac{273+T}{295}\right)$

T420. E Bragg-Gray cavity theory states, in part, that a small air-filled cavity introduced into a phantom (e.g., an ion chamber in a water phantom) will not perturb the passage of photons and secondary electrons in that phantom. The ionization per unit mass of air in the cavity can then be converted into dose, with suitable correction factors. Dose to the air is then converted into dose to the medium.

T421. C A Farmer chamber is a commonly used "thimble" chamber. It is used routinely for calibration of machines.

T422. C Exposure is the charge collected per unit mass of air and the mass of air in the chamber will decrease when the pressure decreases or the temperature increases. Therefore, the charge reading will decrease when the pressure decreases. Humidity has some effect on the reading but is usually less than 1% and is negligible. The temperature would have to be higher in city A than city B to decrease the reading. The stem effect is not relevant to this question.

T423. E The build-up cap is usually made of plastic (low Z) and is used for in-air measurements. Its thickness (plus the thickness of the chamber wall) equals d_{max} for the photon beam in order to provide equilibrium conditions.

T424. D In an air equivalent chamber, a given mass of air is isolated and the charge that is liberated by the radiation is collected. An "air wall" is difficult to achieve but the size of the central electrode and the construction of the chamber wall are adjusted so that the chamber's response with photon energy is similar to the standard air chamber.

T425. C Only the free-air chamber is the absolute exposure standard. Other chambers, TLDs, diodes, and film must be calibrated against a standard chamber.

T426. A The chamber response is affected by air temperature and pressure for an unsealed chamber. The ionization loss by recombination is expected; its amount depends on the chamber design and radiation intensity. W is the average energy required to cause one ionization in the gas. For air, W = 33.85 eV/ion-pair. W is a constant and a correction for the dose determination is not needed.

T427. E The energy of a linac beam as stated by the manufacturer can be misleading. The important parameter is the penetrability of the beam in water.

T428. A Because of the high altitude in Denver, the pressure will be lower. This will reduce the mass of gas in the chamber, giving a lower reading.

TM9.

T429. C The head leakage of high energy accelerators should not exceed 0.1% of the dose rate of the useful beam, both measured one meter from the source.

T430. D

T431. C Also for linear accelerators.

T432. B

T433. E

T434. C The maximum permissible fetal dose is that of the general public, i.e., 0.5 rem/yr = 10 mrem/wk. The maximum head leakage from a cobalt-60 unit is 10 mrem/hr (the average over the whole head must not exceed 2 mrem/hr). Therefore the maximum time is 1 hr. In practice, dose rates received by technologists are far lower, because of the lower average dose rate and increased distance from the head. The best sources of data on typical exposures are the film badge records of the technologists themselves.

THERAPY Dose Measurement

D1. Beam Films

[handwritten: least dose to give acceptable image]

T435.   The most efficient screen to produce a port film with a teletherapy unit is a _____ screen:
  A. high resolution rare earth
  B. long latitude rare earth
  C. par speed calcium tungstate
  D. lead   *[handwritten: ↑ → ↑ spatial + contrast]*
  E. any of the above

[handwritten margin notes: ↑ Z increases the # of photons striking film + # é Compton; hard cassette improves screen; film contrast]

T436.   The contrast of an 8 MV x-ray port film is optimized by:
  A. using ready-pack therapy verification film without a cassette
  B. using a hard cassette with thin lead screens
  C. using a hard cassette with ready-pack therapy verification film
  D. using a high ratio grid cassette to clean up scatter

T437.   Bone-tissue contrast is better in a simulator film than a megavoltage beam film because:
  A. of the type of cassette used
  B. of the type of film used
  C. of the absence of photoelectric interactions in megavoltage beams
  D. of the absence of Compton interactions in simulator beams
  E. of pair production in megavoltage beams

T438.   It is difficult to visualize small bony structures on an 8 MV x-ray "beam" (or "portal") film because:
  A. the Compton process predominates at this energy   *[handwritten: LPE effect]*
  B. at 8 MV, equal masses of bone and tissue will absorb equal numbers of photons
  C. most of the interactions in an 8 MV x-ray beam are independent of Z
  D. scatter cannot be removed by a grid, and this reduces contrast
  E. all of the above   *[handwritten: only useful in diagnostic Z range]*

T439.   The image quality on a cobalt-60 port film is improved if the distance between the film and patient is reduced. The main reason for this is:
  A. geometric source diameter effects are reduced
  B. dose scattered to the film from the patient is reduced
  C. dose scattered to the film from the air space between the patient and the film is reduced
  D. the magnification is reduced

plumber
source size
source → collimator d
collimator to pt of measurement
(depth point)

D2. Patient Monitoring

T440. The type of thermoluminescent dosimetry most commonly used to monitor patient dose in radiotherapy is:

A. calcium sulfate
B. lithium fluoride
C. calcium tungstate
D. sodium iodide
E. calcium fluoride

radiation effects survive meters

T441-443. Match each method of measurement of absorbed dose with the quantity measured:

A. charge Q
B. optical density
C. relative emitted light on heating
D. temperature

T441. Calorimetry   D

T442. Film   B

T443. Thermoluminescent dosimetry   C

T444. Which of the following is **not** true regarding radiographic film used for dosimetry?

A. an advantage is good spatial resolution
B. a disadvantage is increased response to low energy scatter
C. some films have an almost linear dose vs. optical density region
D. anomalous effects can occur when the film is parallel to the beam axis, due to air gaps and misalignment with the phantom surface
E. film is almost tissue equivalent, making it an ideal dosimeter for the energies encountered in radiotherapy

T445. An H-D curve is a plot of:

A. net optical density vs. absorbed dose
B. net optical density vs. photon energy
C. absorbed dose vs. depth in tissue
D. % surviving cells vs. absorbed dose

T446. A diode can be used for all of the following **except**:

A. daily output checks of linacs as part of a QA program
B. measurements of bladder and rectal dose in patients with gynecological applicators
C. measurements of *in vivo* doses during treatment with an appropriate build-up cap
D. calibration of the output of a linac
E. as a probe in a scanning water phantom dosimetry system, where spatial resolution is important

T447.   A film with optical density 1.2 is placed over a film of optical density 1.5. The optical density
         of the combination is:
         A. 0.3
         B. 1.35
         C. 1.5
         D. 1.8
         (E) 2.7

T448.   When x-irradiated crystals of lithium fluoride are heated, they emit:
         A. x-rays
         (B) visible light
         C. electrons
         D. microwaves
         E. none of the above

D3. Units & Definitions

T449.   A sievert is:
         A. 100 ergs per gram
         (B) the SI unit of dose equivalent
         C. dose/RBE
         D. the SI unit of exposure

T450.   The SI units for dose, dose equivalent and radioactivity and their cgs counterparts are:
         A. joule, cGy; sievert, roentgen; becquerel, curie
         B. kerma, cGy; sievert, rem; becquerel, disintegration
         C. gray, erg; thoraeus, rem; disintegration, curie
         (D) gray, rad; sievert, rem; becquerel, curie
         E. roentgen, cGy; rem, sievert; curie, sievert

T451-453. Match the following quantity with its units (units may be used more than once):
         A. joule/kilogram
         B. coulomb/kilogram
         C. joule/second — watt N[/]
         D. eV/ion pair — no value

T451. B Exposure (X)   B

T452. A Gray (Gy)   A

T453. A Sievert (Sv)   A

T454.   The concept of exposure and the roentgen are of limited value in radiation therapy because:
         A. exposure is only defined for photons and not for electrons
         B. the requirements for measuring exposure above 3 MeV are not easily met
         C. for the same exposure in roentgens different tissues will absorb different doses
         (D) all of the above

T455.   For the diagram below, which kerma curve (1 through 4) corresponds with the depth dose curve (thick black line)?
A. curve 1
B. curve 2
C. curve 3
D. curve 4
E. none of the above

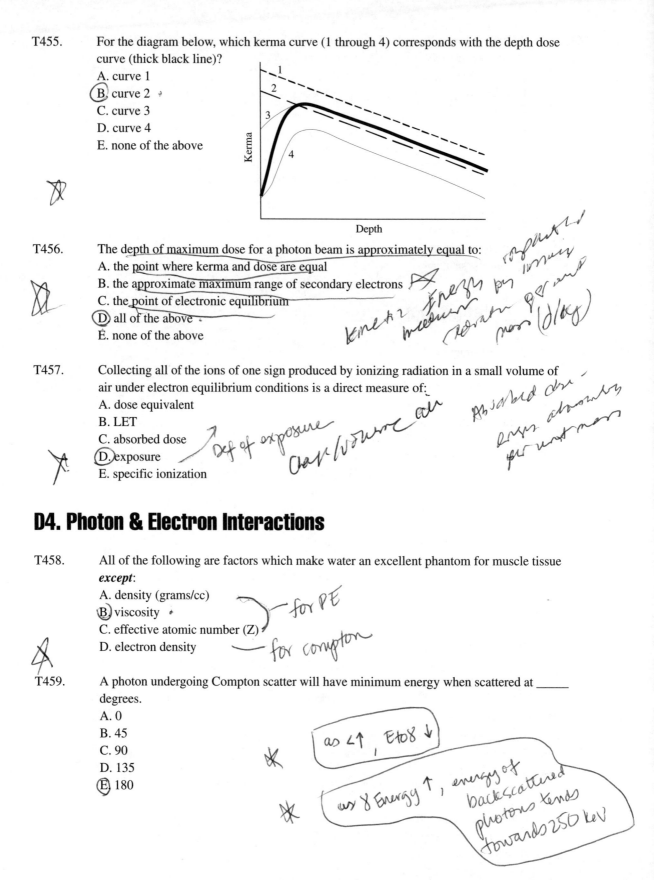

Depth

T456.   The depth of maximum dose for a photon beam is approximately equal to:
A. the point where kerma and dose are equal
B. the approximate maximum range of secondary electrons
C. the point of electronic equilibrium
D. all of the above
E. none of the above

kinetic energy by ionising radiation per unit mass (J/kg)

T457.   Collecting all of the ions of one sign produced by ionizing radiation in a small volume of air under electron equilibrium conditions is a direct measure of:
A. dose equivalent
B. LET
C. absorbed dose
D. exposure
E. specific ionization

Def of exposure   Charge/volume air   Absorbed dose - energy absorbed per unit mass

D4. Photon & Electron Interactions

T458.   All of the following are factors which make water an excellent phantom for muscle tissue *except*:
A. density (grams/cc)
B. viscosity
C. effective atomic number (Z)
D. electron density

} — for PE   — for compton

T459.   A photon undergoing Compton scatter will have minimum energy when scattered at _____ degrees.
A. 0
B. 45
C. 90
D. 135
E. 180

as ∠↑, E to 8 ↓

as γ Energy ↑, energy of backscattered photons tends towards 250 keV

T460.   Regarding photon interactions in water, as beam energy increases from 8 MeV to 25 MeV all
of the following would be expected *except*:
  A. an increase in the % of pair production
  B. a decrease in attenuation per cm of tissue
  C. an increase in PDD
  D. a slight increase in the attenuation per cm of bone   *blc of pair production dependency on Z*
  E. an increase in the % of photoelectric interactions   ✓

T461.   A 2 MeV photon incident upon water will most probably interact by:
  A. photoelectric absorption
  B. Compton scatter   ✓
  C. pair production   *— also possible blc E >1.022 mev*
  D. photonuclear reaction
  E. coherent scatter

T462.   A 30 MeV electron incident upon water may lose energy by:
  A. photoelectric absorption
  B. Compton scatter
  C. pair production
  D. ionization / *excitation*
  E. all of the above

T463.   When a 20 MeV linac x-ray beam interacts with tissue:
  A. pair production predominates
  B. pair production and Compton are about equal   ✓
  C. Compton interactions predominate   *up to 30 mev*
  D. Compton and photoelectric are about equal
  E. Compton, pair production and photonuclear disintegration are about equal

T464.   Stopping power is energy loss by:
  A. electrons per unit pathlength of a medium   ✓
  B. photons per roentgen
  C. electrons per eV absorbed
  D. all of the above

T465.   Which of the following is true about pair production?
  1. It has a threshold of 1.022 MeV
  2. It is Z dependent, and therefore causes more interactions per gram in bone than in tissue
  3. It is the reason why lead is a more effective room shielding material for photons at 20 MV
     than at 10 MV
  4. The number of interactions decreases slowly with increasing photon energy

  A. 1, 2, 3   ✓
  B. 1, 3
  C. 2, 4
  D. 4 only
  E. all of the above

T466. Absorbed dose is lower in bone than in soft tissue when exposed to a cobalt-60 beam because:
A. the f-factor is less for soft tissue
B. the f-factor is less for bone
C. the electron density is less for soft tissue
D. the electron density is less for bone
E. B and D

T467. In a cobalt-60 beam, the predominant mode of photon interaction with tissue is:
A. photoelectric interaction
B. coherent scatter
C. Compton interaction
D. pair production
E. none of the above

T468. When high energy electrons decelerate in a high Z target, the predominant radiation emitted is:
A. gamma rays
B. characteristic x-rays
C. bremsstrahlung
D. lower energy electrons
E. neutrinos

D5. Dose Outside the Treatment Field

T469. The dose to a patient's contralateral breast from tangential breast fields delivering a total dose of 5000 cGy is of the order of:
A. 2500 cGy
B. 250 cGy
C. 25 cGy
D. 5 cGy
E. negligible

T470. Which of the following contribute to the dose received by a patient 10 cm outside the radiation field?
1. head leakage
2. internal patient scatter
3. collimator scatter
A. 2 only
B. 2, 3
C. 1, 2
D. 1, 2, 3
E. 1 only

T471. A pregnant woman is treated for Hodgkin's disease with a mantle field on a 6 MV linear accelerator. The dose to the fetus is approximately what percent of the central axis dose?
A. 25%
B. 10%
C. 5%
D. 0.5%
E. 0.05%

Answers

D1.

T435.   D   A screen absorbs photons and emits electrons, which interact with the film. Lead (Pb) absorbs more high energy photons than the effective materials in the other screens and thus emits more electrons. This is a function of the atomic number (Z) of the screen materials, the thickness of the screens, and the amount of the effective absorbers in the screens.

impt

T436.   B   The interaction between the high energy x-rays in the remnant beam and the Pb screens increases the number of electrons striking the film; a hard cassette improves screen-film contact. Grids are useless in MeV beams as the Pb strips do not attenuate the beam sufficiently. Ready-pack XV film is useful for verification of beam portals, but a localization film gives better contrast.

T437.   C   The probability of PE interactions is proportional to Z cubed, which magnifies the differential attenuation between bone and tissue. Compton, which is Z independent, takes over in the MeV region; differential attenuation here is due only to density differences.

T438.   E   The predominance of the Compton effect in megavoltage photon beams means that there is no differential attenuation on the basis of Z value, as in diagnostic beams, and it is only the difference in density of bone that allows it to be visualized. (Consequently, lungs and sinuses are more easily visualized.) Scatter decreases contrast, but cannot be removed by a conventional grid since the penetrability of the beam would require very great thicknesses of lead and would not be feasible.

T439.   A   Due to the size of the source, rays from the cobalt-60 source converge at the point to be imaged, then diverge as they travel towards the film. Moving the film closer to the patient minimizes this effect, although it increases patient scatter reaching the film, which reduces contrast. Magnification over these distances is not a factor, and air scatter is negligible.

D2.

T440.   B   LiF is ideal for *in vivo* dosimetry for many reasons, notably its linear response over a wide range from mR to hundreds of cGy, the lack of signal fading, which is a problem with other TLD dosimeters, and its relative energy independence.

T441.   D   Calorimetry relates a small increase in temperature delta T in the irradiated volume to absorbed dose. (Delta T produced by 1 Gy = 2.39×10^{-4} °C)

T442.   B   The degree of blackening of the film is measured with a densitometer. This is optical density and is related to dose by an "H-D" curve (a graph of optical density vs. dose).

T443.   C   Thermoluminescence (TL) is the release by heat of light that is stored as energy in a crystal lattice. The TL output is measured and related to a known ratio of TL/absorbed dose.

T444.   E   Because of the high Z of AgBr, film is more sensitive to low energy photons because of photoelectric interactions, and is not tissue equivalent.

T445.   A   A plot of net optical density as a function of radiation dose is termed the sensitometric curve or H-D curve.

T446.   D   Diodes are useful for relative spot-check measurements, and because of the small size of the diode junction region, they can improve accuracy when used as a probe in a water phantom. However, they must be calibrated frequently and their directional dependence must be carefully evaluated before each use. They cannot be used as an absolute dosimeter (e.g., for calibration of the output of a machine).

T447.   E   Optical densities are additive.

T448.   B   The energy absorbed by a TLD (LiF) is stored by electrons trapped in the crystal. When the crystals are heated the electrons drop back to lower energy levels, releasing visible light which is detected and measured by a photomultiplier.

D3.

T449.   B   100 ergs/gram is one cGy. Dose equivalent (Sv) = Dose (Gy) × Q.

T450.   D   The Systeme International d'Unites (SI) has been adopted by the scientific community in place of the cgs system in order to remove numerical prefixes from the units. The unit of dose is the gray (Gy). 1 Gy = 1 J/kg = 100 cGy. The unit of dose equivalent is the sievert (Sv). 1 Sv = 10 rem. The unit of activity (A) is the becquerel (Bq). 1 Bq = 1 dps = $(1/3.7) \times 10^{-10}$ Ci.

thought
1 Sv = 100 rem

T451.   B   1 roentgen = 2.58×10^{-4} c/kg.

T452.   A   1 gray = 1 joule/kilogram.

T453.   A   1 sievert = 1 joule/kilogram. Dose equivalent (Sv) = Dose (Gy) × Q. Q has no units.

T454.   D   The roentgen is only defined for photons less than 3 MeV. Above 3 MeV, it is difficult to measure exposure due to problems of ion recombination, air attenuation and photon scatter. In addition, the unit is not useful in radiation therapy because different tissues will absorb different doses for the same exposure.

T455.   B   From ICRU Report #33, kerma (kinetic energy released in the medium) is the sum of the initial kinetic energies of all the charged ionizing particles (electrons) liberated by uncharged ionizing particles (photons) in a material. Kerma is the energy transferred from photons directly to electrons and is maximum at the surface and decreases with depth since the photon fluence decreases. The kerma curve is lower than the depth dose curve past d_{max}.

T456.   D

T457.   D   This is the definition of exposure.

D4.

T458.   B   An ideal phantom used as a substitute for muscle should have approximately the same density and effective atomic number as muscle so that photoelectric absorption is the same. Electron density should be the same to obtain similar Compton attenuation in the phantom. These three factors are very similar for water and muscle, making water an ideal phantom material. Viscosity is radiologically irrelevant.

T459.   E   As photon energy increases, the energy of backscattered photons tends towards 250 keV. The % of the initial energy given to the scattered photon decreases as the scattering angle increases.

T460.   E   Photoelectric interactions are most likely in the low kV range, but fall off quickly with increasing energy and are absent in the megavoltage region. The increase in pair production explains the slight increase in bone attenuation at 25 MV, as pair production is Z dependent.

T461.   B   Attenuation is the sum of the photoelectric, Compton and pair production attenuations. At 2 MeV, photoelectric absorption is possible but unlikely, Compton attenuation including scatter is most **probable**, pair production is possible because the photon energy is greater than 1.022 MeV but also is unlikely.

T462.   D   An electron interacts mainly by ionization and excitation along its path. (Photoelectric, Compton, and pair production are **photon** interactions with matter.)

T463.   C   Compton interactions are the most likely over the range of (monoenergetic) photon energies from 25 keV to 25 MeV. A 20 MeV linac beam has a spectrum of photon energies with maximum energy of 20 MeV, so pair production will always be less important than Compton. Photonuclear disintegration is very unlikely in tissue at this energy.

T464.   A   (Stopping power is used in the Bragg-Gray cavity formula to calculate dose to a medium from dose to gas in a cavity.)

T465.   A   The number of pair production interactions, unlike Compton and photoelectric interactions, increase as photon energy increases above 1.022 MeV.

T466.   E   The f-factor is the "roentgen to cGy" conversion factor (exposure to absorbed dose). At the cobalt-60 energy range, Compton interactions dominate and the f-factor will vary with #electrons/gram. The number of electrons/gram for bone is slightly less than for tissue, so the f-factor will be lower for bone in the megavoltage range.

T467.   C   Compton interactions predominate over the megavoltage therapy range.

T468.   C   Bremsstrahlung means "braking radiation," and is emitted when electrons are decelerated when passing near the positively charged nucleus. Characteristic x-rays are also emitted but in smaller numbers.

D5.

T469.   B   Assuming a scatter dose of about 5%, the dose would be 250 from the tangentials. Use of a beam splitter will increase the dose and wedges may affect it also. There is controversy over the possible increased incidence of cancer in the contralateral breast as a result of this dose.

T470.   D   Head leakage can have a maximum value of 0.1% of the useful beam, (i.e., 1 mrem per rem), although it may be less than this. Collimator scatter depends on the design of the treatment head and on the distance from the collimator. Patient scatter depends on field size and distance from the field edge.

T471.   D   The dose will be *approximately* 0.5% of the central axis dose due to internal scatter and head leakage, but this depends on distance from field edge to fetus, and the actual value of the head leakage (0.1% maximum).

RP1. Therapy Room Shielding

T472.   NCRP Report #49 gives a formula to calculate the transmission factor "B" which will reduce the primary beam dose to the maximum permissible dose equivalent "P" in a given area. The information needed to calculate "B" is:

$$B = \frac{Pd^2}{WUT}$$

1. Distance from radiation source to area, d
2. Fraction of time area is occupied, T
3. cGy/wk at 1 m, (W workload)
4. Fraction of operating time beam is pointing towards area, U (use factor)
5. HVL of material to be used for shielding barrier
6. Maximum permissible dose in area, "P" (controlled or uncontrolled)

  A. 1, 2, 3, 4, 6
  B. 1, 3, 4, 6
  C. 1, 2, 3, 4, 5
  D. 1, 2, 3, 4, 5, 6
  E. 1, 2, 3, 4

T473.   How many HVLs of concrete are needed to reduce a reading of 75 mrem/hr to 2.5 mrem/hr?
  A. 10
  B. 8
  C. 5
  D. 2
  E. 1

$$\frac{75}{2} \Rightarrow 37.5\big/2 \Rightarrow 18.8\big/2 \Rightarrow \frac{9.4}{2} \Rightarrow 4.7 \Rightarrow 2.35$$
1   2   3   4   2   5

T474.   The maximum energy that must be considered for the design of a primary barrier for a cobalt-60 teletherapy unit is_____ MeV.
  A. 0.255
  B. 0.511
  C. 1.174
  D. 1.332
  E. 2.506

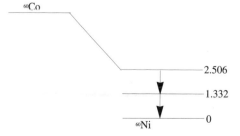

(T)

T475.   The occupancy factor used to calculate the wall thickness required in the direction of a linac
         treatment console: *fraction of working day that area is occupied*
         A. depends on the photon energy
         B. is 1/3 because technologists work 8 hrs out of 24   = 1.0 @ console
         C. depends on the direction of the primary beam
         D. depends on the fraction of time the beam points towards the console
         E. none of the above

T476.   Which of the following would *not* be used in a therapy room shielding calculation?
         A. beam energy
         B. use factor
         C. inverse square law
         D. scatter-air ratio
         E. occupancy factor

T477.   The maximum photon energy that must be considered in the design of a protective barrier at
         right angles to a 30 MeV photon beam is _____ MeV.
         A. 0.255 → @ 180°
         B. 0.511 → annih scatter photon @ 90°
         C. 1.022
         D. 10
         E. 30 → leakage radiation

T478.   In the design of a barrier at 90° to the plane of rotation of the gantry in a cobalt-60 treatment
         room, the maximum photon energy which must be considered is:
         A. 0.5 MeV scattered radiation
         B. 0.25 MeV scattered radiation
         C. 1.17 MeV leakage radiation
         D. 1.33 MeV scattered radiation
         E. 1.33 MeV leakage radiation

T479.   A cobalt-60 source has an output of 100 cGy/min at the isocenter. If the source were stuck in
         the "on" position at 0° gantry, a technologist could stand one meter away from the treatment
         table for about _____ before accumulating 100 mrem.   (1% of beam @ 90°) Key
         A. 1 sec
         B. 10 sec   .001 (100) = .100 rem
         C. 1 min   1 cGy = 1 rad   min
         D. 10 min
         E. 1 hr   100 cGy = 1 rad = 100 rem   100 mrem
                              = 100,000 mrem   min

                                                        (1 minute)

         ↑ .1%
         back ← ⊙ → forward
         ↓ .1%

         x leakage = .1% allowed

T480.   A technologist stands 3 feet from a patient on a linac treatment table. The beam is accidentally turned on long enough to deliver 10 cGy at d_{max} (100 cm SSD) to the patient. The technologist would expect to receive approximately:

3 feet = 1 meter

A. 1 rem
B. 10 rem

10 cGy (.001) = .01 rem

C. 100 mrem
D. 10 mrem

= 10 mrem

E. 1 mrem

Scatter @ 1 meter = 1% of 1 dose

10 cGy = 10 Rad = 10 rem

RP2. Quality Factor

× .001 = 10 mrem   *.01 cGy = .01 rem =*
10 mrem

T481.   According to the NCRP Report #91, the new value assigned to the effective quality factor for neutrons and alpha particles of unknown energy (Q) is:

A. 5
B. 10
C. 15
D. 20
E. 100

RP3. Management of Radioactive Patients

T482.   According to the NRC code of federal regulations #10 Part # 35, a patient with a permanent implant shall not be released from the hospital until:

A. the dose rate from the patient is less than 5 mrem/hr at 1 meter
B. the dose rate from the patient is less than 50 mrem/hr at 1 meter
C. the activity in the patient is less than 30 mCi
D. none of the above

T483.   Safety precautions that should be taken for patients undergoing gynecological implant therapy are all of the following *except*:

A. two implant patients on the same floor should be placed in the same room
B. a "caution radiation" sign should be posted on the patient's door
C. the dose rates in the room and in areas around the patient's room should be surveyed
D. a note should be placed in the chart stating how long a visitor can stay in the room

RP4. Measurement of Occupational Dose Equivalent

T484.   TLD ring badges are worn in addition to a film badge for brachytherapy procedures because:

A. the film badge cannot discriminate between different types and energies of radiation
B. the MPD for the hands is higher than for the whole body and requires a separate measurement
C. the ring badge is simply a back-up for the film badge in case the film is damaged during processing
D. TLD is uniquely sensitive to the particles emitted from brachytherapy sources
E. the skin of the hands has a lower MPD than the whole body

T485.   In a particular hospital, nurses caring for patients with cesium-137 insertions are expected to receive, on average, less than 15 mrem/wk. Which of the following is true?
        A. the nurses must be defined as radiation workers and must be monitored with bi-weekly film badges
        B. they require no monitoring as they will receive only about 1/2 of the MPD for radiation workers
        C. they receive less than 1/4 of the MPD for radiation workers, and although monitoring is not therefore legally required, it is prudent in order to ensure (and demonstrate) good radiation safety technique
        D. wearing lead aprons would reduce the nurses' exposure by a factor of 10
        E. C and D only

T486.   Therapy technologists' exposure is not measured with routine blood counts because:
        A. the cost would be prohibitive
        B. other factors affect blood count, and would mask the effect of radiation
        C. the dose required for a measurable effect would be far greater than the MPD
        D. the turn-around time for the test would be too long to be of any practical use
        E. neutron dose could not be measured this way

T487.   Which of the following is true about film badges for personnel monitoring?
        A. they can measure alpha exposure
        B. they can measure dose accurately
        C. they can measure exposure accurately
        D. they can differentiate between beta and gamma radiation
        E. none of the above are true

RP5. Dose Equivalent Limits

T488.   NCRP Report #91 recommends the following dose limits for:
        (a) Annual individual non-occupational exposure (infrequent) and (b) Total dose to an embryo
        A. 0.5 rem and 0.5 rem
        B. 0.5 rem and 0.1 rem
        C. 0.1 rem and 0.5 rem
        D. 0.1 rem and 0.1 rem
        E. 0.5 rem and 0.05 rem

T489.   NCRP Report #91 recommends that the annual occupational exposure for the lens of the eye should not exceed:
        A. 1 mSv
        B. 5 mSv
        C. 10 mSv
        D. 150 mSv
        E. 500 mSv

RP6. Regulations

T490.   The NRC requires that sealed radioactive sources be leak-tested:
A. initially (when received)
B. initially then monthly
C. initially then every 6 months
D. initially then annually
E. annually

T491.   The NRC requires which of the following quality assurance measures for brachytherapy sources?
1. quarterly inventory
2. a log showing the location of all sealed sources removed from the safe
3. annual training in safe handling of sources for all personnel who care for brachytherapy patients
4. a leak-test of all sealed sources every 2 years *(every 6 months)*

A. 1, 2, 3
B. 1, 3
C. 2, 4
D. 4 only
E. all of the above

RP7. Detectors

T492.   When looking for a dropped iodine-125 seed on the floor of the operating room, a Geiger counter would be used rather than an ionization chamber survey meter because:
A. it responds to low dose rates because of gas multiplication
B. it converts an interaction in the detector into an electrical pulse } *also true of survey meters*
C. it can be calibrated for the energy of the seeds
D. it is a light, portable instrument
E. none of the above

T493.   An iodine-125 seed is dropped on the floor of the operating room during an implant. The best way to locate the seed is with:
A. a survey meter
B. a thimble ionization chamber connected to an electrometer
Na I — C. a portable scintillation detector *or a Geiger => both provide the necessary amplification*
D. a pocket dosimeter
E. any of the above

T494.   Which instrument is **best** suited for finding a lost cesium-137 source?
         A. survey meter
         B. ion chamber
         C. TLD
         D. Geiger counter   ⟹ uses gas amplification
         E. diode

T495-498. Select the detector most appropriate for the use specified:
           A. large volume ion chamber survey meter
           B. BF_3 counter
           C. thimble ionization chamber
           D. pocket dosimeter
           E. well counter

T495.   E Counting swabs used to leak-test brachytherapy sources

T496.   C Calibrating the output of a 10 MV linac C   Farmer <1cc ion chamber

T497.   B Measuring neutron leakage around a linac head   BF3 (B = Boron = neutron capture)

T498.   A Measuring photon leakage around the head of a cobalt-60 unit — large volume, s/c low dose rate

RP8. Miscellaneous

T499.   A standard 0.5 mm Pb apron is used to cover the gonads of a patient receiving
         cobalt-60 teletherapy to the pelvic area. As a result, the:
         A. gonadal dose is significantly reduced
         B. genetic dose is significantly reduced
         C. risk of sterility is significantly reduced
         D. skin sparing is lost
         E. depth dose improves due to a decrease in scatter

T500.   Implantation of the tongue by iridium-192 seeds results in less exposure to the radiotherapist
         than a radium needle implant because:
         A. iridium-192 has lower energy gammas than radium
         B. of the difference in the exposure rate constants
         C. of the difference in the RBEs
         D. iridium-192 is an afterloading procedure
         E. iridium-192 has no radioactive daughters

T501.   Potassium iodide is useful:
         A. to block the uptake of radioactive iodine released by a nuclear accident
         B. to protect the thyroid of a patient about to undergo a thyroid scan with ^{99m}Tc
         C. to block the uptake of iodine-131 in normal thyroid for patients undergoing ablative therapy
            of the thyroid
         D. to treat patients with cancer of the thyroid
         E. none of the above

Answers

RP1.

T472.   A   The equation for the transmission factor "B" is: $B = P d^2 / WUT$. The HVL of the barrier material will be needed to calculate the barrier thickness, but is not required to calculate B.

T473.   C   It will take 5 HVLs of concrete or any other material to reduce 75 to 2.5.

T474.   D   Cobalt-60 emits two γ-rays of 1.33 and 1.17 MeV. The maximum photon energy is 1.33 MeV.

T475.   E   The occupancy factor depends only on the fraction of the working day that the area could be occupied; it is always 1.0 for treatment consoles.

T476.   D   SAR is used in Clarkson irregular field calculations.

T477.   E   The design of a protective barrier must include contributions from primary, scattered and leakage radiation. In this case, the primary radiation is not a factor and the maximum scattered radiation is only about 0.5 MeV. However, the leakage radiation through the tube housing is the same energy as the primary, 30 MeV.

T478.   E   At 90° the maximum energy of scattered radiation is 0.5 MeV; however, head leakage will have a maximum energy of 1.33 MeV.

T479.   C   90° scatter is about 0.1% of the primary beam, so the dose rate lateral to the couch would be about 100 mrem/min, neglecting inverse square considerations.

T480.   D   The scatter at 1 meter is about 0.1% of the primary dose, or 1 mrem per cGy at the isocenter.

RP2.

T481.   D   Q has changed from 10 to 20 based on "the information now available on the RBE for neutrons for a variety of cellular effects *in vitro*" — see NCRP 91, p. 12.

RP3.

T482.	A	One requirement for release of a patient with a permanent implant is that the dose rate at 1 meter from the patient is less than 5 mrem/hr.
T483.	A	The patient should not be placed in the same room as another patient.

RP4.

T484.	B	The hands will receive a higher dose than the body in a brachytherapy procedure and should be monitored separately. The current recommended MPD for the hands is 75 rem (750 mSv) per year.
T485.	C	Lead aprons contain far too little lead to be effective against cesium-137 gammas (660 keV, HVL 0.6 cm Pb). Monitoring is not legally required if the dose is not expected to exceed 1/4 of the MPD or 25 mrem/week. However, accidents are always possible with afterloading devices, and a film badge would monitor dose received in this event. In routine use the low readings would serve to reassure the nurses.
T486.	C	About 25 cGy is required for an observable effect.
T487.	D	Film badges have various types of filters in order to obtain some information about the beam quality. However, film has an energy dependence and cannot measure dose or exposure very accurately.

RP5.

T488.	A	See NCRP Report #91.
T489.	D	

RP6.

T490.	C	The NRC requires that sealed radioactive sources must be leak-tested initially then every 6 months if (1) they contain more than 100 microcuries of beta or gamma emitting material or more than 10 microcuries of an alpha emitter, (2) they have a half-life greater than 30 days, and (3) are in any form other than a gas.
T491.	A	Sealed sources must be leak-tested every 6 months.

RP7.

T492. A B-D are also true of survey meters. In order to detect the weak signal from a single seed at some distance, we require a detector which multiplies the signal. A calibrated meter is not necessary for detection only.

T493. C A typical 0.5 mCi iodine-125 seed emitting 27 and 35 keV x-rays will cause a negligibly small ionization current in either A or B. It can be detected by either a scintillation or Geiger counter, both of which provide the necessary amplification of the small signal. D is designed to integrate dose and cannot be used as a dose rate meter.

T494. D The Geiger counter is most useful for detecting the presence of radiation. This is because a small signal from a low dose rate source is magnified by gas amplification.

T495. E

T496. C

T497. B

T498. A

RP8.

T499. D A 0.5 mm Pb apron will absorb approximately 1% of a cobalt-60 beam. At the same time, it will considerably increase the scattered radiation to the patient and remove the skin sparing effect.

T500. D Afterloading systems always deliver far less dose to staff than implanting active sources in the operating room.

T501. A Potassium iodide is taken up by the thyroid, reducing the subsequent uptake of radioactive iodine, e.g., iodine-131.